Land of Plants in Motion

Perspectives on the Global Past

Anand A. Yang and Kieko Matteson
SERIES EDITORS

Land of Plants in Motion

Japanese Botany and the World

Thomas R. H. Havens

University of Hawai'i Press
Honolulu

Paperback edition 2021

Printed in the United States of America

26 25 24 23 22 21 6 5 4 3 2 1

Library of Congress Cataloging-in-Publication Data

Names: Havens, Thomas R. H., author.
Title: Land of plants in motion : Japanese botany and the world / Thomas R. H. Havens.
Other titles: Perspectives on the global past.
Description: Honolulu : University of Hawai'i Press, 2020. | Series: Perspectives on the global past | Includes bibliographical references and index.
Identifiers: LCCN 2020007859 | ISBN 9780824882891 (cloth) | ISBN 9780824883447 (pdf) | ISBN 9780824883454 (epub) | ISBN 9780824883461 (kindle edition)
Subjects: LCSH: Botany—Japan—History. | Plants—Japan—History. | Plants—Migration—History.
Classification: LCC SB87.J3 H38 2020 | DDC 581.952—dc23
LC record available at https://lccn.loc.gov/2020007859

ISBN 978-0-8248-8973-9 (pbk.)

Cover art: (*Top*) Triple hemispheres of the Yumenoshima Tropical Greenhouse Dome. (*Bottom*): Akatsuka Botanical Garden, Tokyo. Photos by author.

University of Hawai'i Press books are printed on acid-free paper and meet the guidelines for permanence and durability of the Council on Library Resources.

In memoriam

Marius B. Jansen (1922–2000)

Delmer M. Brown (1909–2011)

Contents

Preface

The inspiration to study the history of Japanese plants arose from several sources: growing up near the fields and forests of Pennsylvania, studying at two university campuses designed in part by the American landscape architect Beatrix Farrand (1872–1959), living next to the Connecticut College Arboretum and later near Boston's Arnold Arboretum—and most of all visiting parks, woodlands, and botanical gardens during various research trips to Japan since 1960. *Land of Plants in Motion* explains the long evolution of plant life in Japan and East Asia, the spread of Japanese botanical species worldwide in geological and historical time, and the emergence of plant sciences in Japan since the mid-nineteenth century. It provides context for widely held sentiments about nature expressed by Crown Prince Naruhito (now the Reiwa emperor) in November 2018: "I hope a culture of growing forests, using trees in our lives and coexisting with trees, will spread widely from here in Tokyo to the whole of Japan, and to the next generation," despite ceaseless urbanization and people's ever more virtual, not actual, experience of the natural world.[1]

This book follows the Japanese practice of placing family names before personal names, except for citations of Western-language publications. Romanizations of Japanese terms follow the modified Hepburn system found in *Kenkyusha's New Japanese-English Dictionary,* fifth edition (Tokyo: Kenkyūsha, 2003). Macrons are omitted over long vowels in the most widely known place names. Scientific names of plants found in historical sources sometimes differ across time. Unless otherwise indicated, photographs are my own. A visiting scholar appointment at Meiji University and funds from Northeastern University speeded my studies; I'm grateful to these sources for indispensable assistance. Hashimoto Seiko and Hashimoto Minoru have graciously encouraged my research throughout my career. I'm also grateful to many colleagues in conversation and correspondence, especially Robert A. Askins, Azuma Yoshirō, Bruce Batten, Beverly J. Bossler, Edward I. Brodkin, Lawrence Buell, David E. Boufford, Martha Butler, Sara A. Butler, Chō Kyō, Dairokuno Kōsaku, Glenn Dreyer, Benjamin C. Duke, William (Ned) Friedman, Kanchi N. Gandhi, Sheldon Garon, Larissa E. Glasser, Andrew Gordon, Junko Habu, Keven Halliday, Joanna Handlin Smith, Valerie Hansen, Hiwatari Nobuhiro,

David L. Howell, Itō Taiichi, William W. Kelly, Steven R. Manchester, Ian J. Miller, Gregory W. Noble, Ohba Hideaki, Christopher M. Parsons, Lisa E. Pearson, Frederik L. Schodt, David H. Slater, Kent C. Smith, Margaret C. L. W. Smith, C. Sarah Soh, Heather Streets-Salter, Sally L. Taylor, Philip Thai, Conrad D. Totman, Stephanie Turnbull, Ezra F. Vogel, Richard von Glahn, Gretchen Wade, Judith A. Warnement, R. Scott Warren, Dennis C. Washburn, Ikuko Watanabe Washburn, and Samuel H. Yamashita. I'm especially grateful to Curtis C. Daeler and two anonymous Press readers for careful readings of the manuscript. I'm also greatly indebted to Masako Ikeda, Grace Wen, Debra Tang, Ivo Fravashi, and the staff of the University of Hawai'i Press for expert editorial and production care.

Land of Plants in Motion is dedicated to the memory of two sterling historians who set lofty standards of scholarly excellence and personal kindness: Marius B. Jansen (1922–2000) and Delmer M. Brown (1909–2011). Both expertly guided me in becoming a professional teacher, researcher, and campus citizen by instilling in me a deep respect for the responsibility scholars have in the education of future generations, by example as well as by precept. I'm grateful to my children, William Hamilton Havens, Carolyn Havens Niemann, and Katherine Havens Whitten, as well as to the rest of my family for love and support: my sister Anne Havens Fuller, my daughter-in-law Julie Hunt, son-in-law Michael Niemann, and grandchildren Adam Niemann and his wife Panayiota Kuvetakis, Jacob Niemann, Matthew Niemann, Emily Whitten, and Nate Whitten, as well as my in-laws Nors Josephson and the Thornbers: Carol, Jay, Karvel, Kat, and Nora. Above all, I'm grateful to my wife Karen L. Thornber, an expert linguist and leading scholar of Asian and global literatures and cultures with a deep knowledge of Japanese and Asian ecology. I'm profoundly thankful for her stimulating intellect, endless proofreading, and love, learning, and laughter beyond measure or description.

Land of Plants in Motion

Introduction

"Almost 80 percent of plants used in American landscaping are of Japanese origin,"[1] wrote John L. Creech (1920–2009), former director of the U.S. National Arboretum, in 1994. "A vast number of plants cultivated today in American gardens owe their origins to the wild and cultivated plants of Japan."[2] Creech was right that plants from Japan have spread widely to Europe, the Americas, and elsewhere since at least the 1680s, when the German naturalists Andreas Cleyer, George Meister, and Engelbert Kaempfer, representing the Dutch East India Company (VOC) at Nagasaki, sent specimens of Japanese vegetation back home. At first, rich Europeans sought new and exotic floral species to adorn their estates. Soon European nurseries and scholars began importing live plants, seeds, and herbarium specimens from Japan, prompted by scientific curiosity and imperial ambition (Christopher Hodgson, a British consul, said in 1861, "There are as fine oaks in Yezo [Hokkaido] as in any part of the world . . . of such prodigious trunks, and such apparently sound wood, that they are fit for the construction of any man-of-war in the world"[3]). An azalea mania in the eighteenth century, a fern fad in the 1860s, a bonsai boom starting in 1879, and a fervor to construct "Japanese" gardens in the first decade of the twentieth century led successive generations of Europeans to import plants from Japan, as was true of American horticulturalists in Philadelphia, New York, and Brookline, Massachusetts after the U.S. Civil War. A few imports, such as kudzu and honeysuckle, were eventually deemed invasive, even if not menacing in Japan—because in North America they outcompete native plants, even though in Japan they don't competitively replace other plant species.

But what does it mean to call these plants "Japanese"? Today's Japanese islands clung to the eastern coast of the Eurasian continent until 20 million years ago and partly to the Korean peninsula until 2 million years ago; land bridges to the continent existed on and off during various ice ages until as recently as 12,000 years ago,[4] permitting plant exchanges in both directions. Once the land bridges were submerged by rising seas, new species evolved in Japan now that it was isolated from the continent. The eastern Himalaya, western China, southern Korea, and Japan comprise an East Asian floristic zone dating to more than 100 million years ago; ancestors of many trees and other plants now found in Japan spread there from China in Tertiary times (66–2.6 million years ago) if not before. Japan was little affected by glaciation during ice cycles, allowing for considerable continuity of vegetation, much of it stemming from early continental roots. In 1948 Harvard botanist Elmer Drew Merrill (1876–1956) foreshadowed Creech's remarks on horticultural imports, writing that "China is known as the mother of gardens. A surprisingly high percentage of our cultivated ornamental plants originated in that country"[5]—and that such plants often reached the Americas and Europe by way of Japan. Japanese phytogeographer Hara Hiroshi (1911–1986) conducted five surveys in the eastern Himalaya from 1960 to 1972 to identify the continental forebears of Japanese botanical life, analyzing evolution from common ancestors in various plant groups, today widely separated in the two regions.[6] One estimate suggests that 40 percent of the fern species currently found in Guizhou province, China, are shared with Japan, likely spread by spores carried across the seas by winds.[7]

In geological timescales, plant distribution occurred naturally, but in the historical period, plants were often introduced by people. Buddhism's arrival in 538 CE also brought medicinal plants such as tea and chrysanthemums from China and Korea. The Japanese apricot, bearing red flowers, was introduced from China, joining preexisting wild individuals in Kyushu.[8] The ginkgo, a robust street tree that symbolizes the Tokyo Metropolitan Government, came from mountains south of the Yangzi river. Today China, South Korea, and Japan dispute the true home of cherry trees, but they likely originated in the eastern Himalaya and arrived in Japan during China's Tang dynasty (618–907).[9] The lotus, peony, camellia, and numerous types of trees came to Japan from China many centuries ago; in the Edo era (1600–1868), orchids, various roses and magnolias, Chinese redbud, *Clematis florida,* trident maples, and many others found new homes in Japan. More recently, Himalayan cedars were established via seeds starting in 1879, followed in 1948 by the *Metasequoia,* a genus at least 150 million years old that was once widely distributed in northern temperate regions but was not identified until 1941 by Miki Shigeru (1901–1974), a Japanese paleobotanist working from fossil samples. In 1944 Chinese foresters discovered a few living groves of this giant tree in Sichuan and Hubei provinces and confirmed them as metasequoias in 1946. Handsome descendants grown from seeds collected in China are now found in parks throughout Japan and around the world.[10]

While acknowledging that as many as two thirds of extant plant species in Japan originated on the Asian continent, botanists currently regard 1,860 terrestrial species as native to Japan, with perhaps 110 genera also having close

Figure I.1 Metasequoia (*Metasequoia glyptostroboides*), 2017. National Museum of Nature and Science Tsukuba Botanical Garden, Ibaraki prefecture.

affinities with counterparts in the southern Appalachian mountains of North America.[11] Even some plants introduced by people, such as bamboo, that came to Japan from China are by now considered native, hence "Japanese." Floral expert Tsukamoto Yōtarō (1912–2005) has explained, "It is fair to say that no other modern nation has developed such a special interest in its gardens based entirely on native plants, save for those plants introduced from China so many centuries past. The Chinese plants have been in gardens for so long, and are so ubiquitous, that they are thought by most Japanese to be indigenous."[12] In this way, specialists consider many naturalized plant species now found in Japan as "Japanese," based on length of time in the country as well as diachronic differentiation from their close relatives abroad.

Land of Plants in Motion: Japanese Botany and the World is the first study to bring together two companion stories, both of which are relevant for the history of Japan, the history of scientific investigation, and global history. The first details how the Japanese islands were initially incorporated into the East Asian floristic zone, and how the ancestors of many Japanese plants were also able to spread from this zone to eastern North America by 5 million years ago. As a result of climatic and environmental factors, the Japanese islands developed an astonishing wealth of plant species. The geohistorical background provided by these developments that occurred in geologic time is crucial to understanding the second story, which centers around the rise of Japanese botanical sciences since 1600, beginning as the study of herbal medicine and natural history but growing into the field of modern plant science in the later nineteenth century. Given their country's rich endowment of flora, since 1860 Japanese botanists and amateur aficionados have devoted increasingly specialized attention to the scientific study of plants and the creation of botanical gardens. Indeed, few if any peoples match the Japanese in their knowledge and appreciation of the plant life surrounding them. These same plants have been widely sought by Europeans since the late seventeenth century and by North Americans since the nineteenth century as embellishments of their ornamental gardens and as objects of scientific study.

This Introduction outlines the chapters that follow, sketches the flora in Japan's five current vegetation zones as mise-en-scène, and defines *native, naturalized,* and *invasive* species as these terms are used by scholars and lay persons. Chapter 1 recounts the evolution of plant life in the East Asian floristic zone from the late Cretaceous period (100–66 million years ago) through the Tertiary (66–2.6 million years before the present) and Quaternary (2.58 million years ago to the present) periods, with focus on the ancestors of plants found in Japan today. This chapter also analyzes the nineteenth-century recognition that the flora of eastern North America contains many of the same plant groups as Japan at the genus level, although species often differ; the similarities suggest descent from a common flora once found in both regions rather than coincidental evolution.[13]

Chapter 2 turns to the arrival of medicinal plants (Ch. *bencao,* Jp. *honzō*) from China in the sixth century and the intensive study of herbal medicine in Japan during the Edo era (1600–1868). Successive employees of the Dutch East India Company at Nagasaki, such as Engelbert Kaempfer (1651–1716), Carl Peter Thunberg (1743–1828), and Philipp Franz von Siebold (1796–1866), brought Western knowledge of biology to Japan and made Japanese flora known to Europe via plant specimens and publications. A number of Japanese naturalists, some of whom trained at the Dutch enclave in Nagasaki, moved beyond the study of herbal medicine in the later Edo period to lay the foundations for modern botany in Japan in parallel with but largely independent of their European peers.

The third chapter takes up the export of Japanese plants to Europe and North America starting with the late eighteenth century and accelerating in the 1860s at the hands of intrepid plant collectors such as Robert Fortune (1812–1880) from Scotland, the Russian Carl Johann Maximowicz (1827–1891), and John Gould Veitch (1839–1870) of London. By 1891 the Western penchant for trees and ornamentals from Japan was so great that Suzuki Uhei (1834–1910) and his son Hamakichi (d. 1925) founded the Yokohama Nursery Company, the first Japanese-owned firm to export indigenous vegetation to nurseries abroad.[14] Competitors soon emerged, inter alia to service a boom in "Japanese" gardens in Britain in the 1910s and 1920s.[15] Plant quarantines imposed to forestall possible viruses slowed live exports to the United States in 1912, 1919, and the 1970s, but seeds from Japan continue to be distributed globally.

The final three chapters focus on the development of plant biology in Japan from the 1870s to the present, highlighting the growth of one of the country's most vibrant areas of scholarly research. Chapter 4 examines botanical research and instruction at the University of Tokyo (renamed Imperial University in 1886, Tokyo Imperial University in 1897, and University of Tokyo in 1947), pioneered by Itō Keisuke (1803–1901) and three specialists in taxonomy and morphology: Cornell alumnus Yatabe Ryōkichi (1851–1899), Matsumura Jinzō (1856–1928), and Miyoshi Manabu (1862–1939). Yatabe, Matsumura, and Miyoshi served in succession as director of the university's renowned botanical garden at Koishikawa. Sapporo Agricultural College (est. 1876, renamed Hokkaido Imperial University in 1918 and Hokkaido University in 1947) and Kyoto Imperial University (est. 1897, renamed Kyoto University in 1949) became other leading research centers. Scholarly societies for plant research were first established in the 1880s, publishing journals that continue today, while a number of botanical gardens and arboretums were founded at the beginning of the twentieth century. These and other pioneers enjoyed worldwide acclaim for their scholarship and helped earn their country a strong reputation for excellence in biological research.

Chapter 5 analyzes the elaboration of botanical knowledge during Japan's age of empire, 1905–1945, framed by (1) further specialization and diversification among Japanese botanists at home, and (2) the seldom-considered topic of Japanese colonial botany (the study, naming, cultivation, and marketing of plants in colonial context).[16] An especially prominent scholar whose work focused on the main Japanese islands was Makino Tomitarō (1862–1957), who quit primary school and never earned a university degree, became Japan's greatest popularizer of botanical knowledge, and eventually served as a lecturer at the University of Tokyo. Another self-taught specialist in the same generation was Minakata Kumagusu (1867–1941), a naturalist and ethnologist as well as researcher on slime molds who became one of Japan's first ecologists.[17] Miyoshi Manabu introduced ecology to Japan in 1895, but it became an independent field of research only in the 1930s. Other scholars stepped beyond the foundational subjects of taxonomy and morphology to explore newer branches of plant science, such as cytology and genetics. Shibata Keita (1877–1949) published Japan's first monograph on photosynthesis in 1931,[18] while major studies began to appear on grasslands, mosses, lichens, and marine algae. During the 1920s and 1930s positions in botany were launched at Kyoto, Kyushu, Tohoku, and Tokyo Bunrika Universities. Meanwhile specialists on plants in Japan's overseas possessions fanned out to colonies in Karafuto (southern Sakhalin island), Korea, and Taiwan, as well as to Japanese-occupied areas in Manchuria, Mongolia, and the Pacific islands. When Taihoku Imperial University (now National Taiwan University) was established in 1928, three chairs in botany with focus on tropical plants were quickly added; in colonial Korea, Japanese researchers on economic plants conducted detailed studies to improve crop yields, chiefly to the benefit of the metropole rather than the colonized peoples. Throughout the era of imperialist expansion, Japan's particular approach to colonial botany was one of plant collection, not exploration.

The sixth and final chapter analyzes the intense preoccupation with Japanese plant life by scholars in that country and horticulturalists worldwide since 1945. Researchers have produced robust studies of discrete genera, including dendrology, economic/useful plants, woody and nonvascular taxa, ferns, mosses, lichens, seaweeds, and tropical species. By some accounts morphology—the study of plant forms and structures—languished after World War Two, while research on botanical hormones and the movements of plants accelerated greatly. A National Genetics Institute, founded in 1949, helped introduce plant genetics four years later.[19] Sixty-six institutions created under the new university system of 1949 gradually added instruction and research in botany, a huge expansion from prewar times. Experts in molecular ecology focused on natural plant communities and in the 1970s studied the production of plant material under the severe environmental conditions of that decade.[20] New research on DNA starting in the 1980s allowed scholars to clarify biological lineages, so that "plant classification truly faced revolutionary times"[21] as many species were reassigned to different families. The Japan Association of Botanical Gardens, founded in 1947, today counts more than 150 members; most were established since the end of World War Two, reflecting both public and scientific interest in studying plant life.[22] Japan also has 34 national parks, 56 quasinational parks, and many hundreds of city parks for various levels of nature study.

Land of Plants in Motion closes with an afterword on Japanese botany and the world, including an analysis of ecological threats faced by Japan today. Early in the twenty-first century the combined toil of four hundred government specialists, university scholars, and citizen volunteers identified 1,665 endangered plant species, and more that were threatened or extinct, showing that the environmental peril faced by Japanese vegetation "was graver than had been feared."[23] Japan continues to import floral species from around the world, as well as contributing live, dried, and fossil specimens to researchers, horticulturalists, and gardeners in more than three dozen countries.

From Alpine to Subtropical

Japan today displays an astonishing variety of vegetation, from alpine regions in Hokkaido and northern Honshu to near-tropical conditions in lower Kyushu and Okinawa. How best to grasp this diversity calls for appropriate terminology. Microbiologist Carl R. Woese and colleagues proposed in 1977 that there were three domains of life: archaea (oldest single-cell organisms), bacteria, and eukarya (all organisms with cell nuclei, including amoebas, fungi, animals, and plants).[24] This tripartite division, although initially greeted with skepticism, is now widely accepted by specialists in paleogeology and molecular evolution.[25] Because *botany* is sometimes understood to mean just green plants, a label preferred by some specialists is *the study of algae, fungi, and plants.*[26] A more comprehensive, if self-referential, view is that a plant is "any organism traditionally studied by botanists," including fossil specimens and both photosynthetic and nonphotosynthetic groups.[27] More recently a number of scholars have preferred the terms *plant biology* or *plant science* in lieu of *botany.* Nature comes in all colors, but my focus is mainly on higher plants of relatively complex structure, including trees and other vascular plants, which are almost always green. Much of the prehistoric human past was an age of wood, when trees and shrubs dominated food chains and material culture.[28] Although relics of plant materials are scarcer than stone or iron, archaeologists since the 1990s have used plant remains to understand how humans adapted to their socioenvironmental contexts.[29] With its relative lack of ice-age glaciation, Japan is ripe soil for such investigations.

Forests cover 30 percent of earth's land surface and account for half its terrestrial plant life; two thirds of Japan's land area is forested, supporting as much as 90 percent of the country's plant activity.[30] Dendrologist Kira Tatsuo (1919–2011) has asserted, "Japan is unique in all the world in its uninterrupted forests from the subarctic to the subtropical."[31] Since life arose on earth nearly 4 billion years ago, roughly 5 billion species of plants, animals, and other life forms have emerged, 99 percent of them now extinct; estimates of species currently alive range from 5 to 100 million, not including bacteria and viruses.[32]

Excluding algae, ferns, molds, and fungi, between 298,000 and 380,000 plant species inhabit the planet's oceans and landforms, of which Japan counts more than 7,000 vascular land species, nine tenths of them angiosperms (flowering plants)[33]—probably the world's richest trove of woody flora in the north temperate region.[34] Japan's forested areas contain about 600 tree species and many times that number of other wild plant species.[35] The country also counts 500 genera of economic plants—once patriotically defined for the American occupation government as "plants useful to man in Japan,"[36] mainly foodstuffs (useful) and construction materials (economic).

Japan's climate zones can be parsed both by altitude, from sea level to mountains 3,000 meters high, and by latitude, from 25 to 45 degrees north. Plants in Honshu can be further divided between the Pacific and Japan sea sides of the island; highland areas facing the Japan sea with the heaviest snows tolerate only dwarf bamboos and random conifers. Analyzed by vertical distribution, lowlands up to 500 meters above sea level in much of the country comprise a warm temperate zone, filled with broadleaf evergreens and flowering shrubs; regions between 500 and 1,500 meters form a cool temperate zone, with a mix of deciduous and conifer trees. The next 1,000 meters of altitude constitute a subalpine zone with a thick canopy of conifers and shade-resistant plants on the forest floor. Above 2,500 meters is a frigid alpine zone with grasses, lichens, hardy dwarf plants, and low creeping pines.[37]

Examined from north to south, alpine meadows are most common in the mountains of Hokkaido and on the tops of the tallest mountains in Honshu; many plants found there moved south from eastern Siberia during ice ages, including dwarf or creeping pines (*Pinus pumila*) that can live as long as two hundred years.[38] Summer wildflowers provide a sparkling palette: meadow buttercups (*Ranunculus acris*), narcissus anemones (*Anemone narcissiflora*), highly fragile bleeding hearts (*Dicentra peregrina*) with hanging pink flowers, and bright violets (*Viola alliariifolia*) found only on mount Daisetsu in Hokkaido. Alpine terrain also hosts many small azalea (*tsutsuji*) shrubs, hemlocks (*Tsuga zakura*) in dry meadows, and in craggy or windy places Kamchatka globeflowers (*Trollius riederianus*) with lemon-yellow blooms.[39] Alpine specimens were bought to Japan's cities in the 1830s when the quest for medicinal plants accelerated; in the early twentieth century alpine plants were cultivated in pots and occasionally exhibited by amateur flower lovers in Tokyo.[40] Today adventurous backpackers and day-trippers throng Japan's national parks to experience alpine vegetation firsthand and discover landscapes scarcely seen elsewhere.

Subalpine districts in the lower elevations of Hokkaido and on the upper mountainsides in Honshu are replete with conifers, some of which have made their way south from the Eurasian taiga, the world's largest land biome. Japan's evergreen conifers have had to adjust to high humidity and sometimes wet areas with mossy forest floors. The year-round gloominess of evergreen conifer regions means that forest floor plants must adapt to low light and strong acidity. Some orchids, poppies, fungi, and semiparasitic plants grow well in these conditions. Sakhalin firs (*Abies sachalinensis*) thrive in northern Hokkaido, as do Japanese spruces (*Picea jezoensis*) and Marie's firs (*Abies mariesii*) in upland Honshu. Deciduous conifers such as Japanese larches (*Larix kaempferi*) are also widely distributed. Erman's birches (*Betula ermanii*) with creamy

Figure I.2 The upper Ara river in Chichibu-Tama-Kai national park cuts through thick forests northwest of Tokyo to spill out into Tokyo bay just east of central Tokyo, 2009. Courtesy Karen L. Thornber.

white bark are scattered in subalpine evergreen areas, as are white birches (*Betula tauschii*), a species also found in Finland. The seeds of white birches (*shirakaba*) spread widely and shoots sprout quickly, advancing into open areas cleared by fires or foresters. In southern parts of the subalpine zone, Clinton's lilies (*Clintonia borealis*) and species of the *Chrysanthemum* genus are well established. Many wet meadows and high-altitude boggy marshlands caused by melted snow contain Amur daylilies (*Hemerocallis middendorffii*), species of the *Hosta* and *Sphagnum* genera, and Asian skunk cabbage (*Lysichiton camtschatcensis*). One of the best-known subalpine wetlands is Osegahara in Honshu, now much compromised by tourism and an electricity generating station. The sandy coast of Hokkaido, with its onshore breezes and salt sprays, features ancient flower fields of daylilies and other broadleaf plants that blossom only briefly and seldom revive, such as those at Abashiri, after being trampled by human footfalls.[41]

Japan's cool temperate zone forests contain huge ranges of fir, maple, birch, and spruce genera, as well as great oaks (*Quercus crispula*), Japanese hemlocks (*Tsuga sieboldii*), cryptomeria (*Cryptomeria japonica*), hinoki cypress (*Chamaecyparis obtusa*) in rocky areas, various species of maples (*Acer*), and the world's largest beech (*Fagus crenata*) forest in the Shirakami mountains of northern Honshu.[42] Beeches on the shore of the Japan sea coast have broader leaves than those facing the Pacific, presumably because of a shorter growing season and the need to absorb snowmelt.[43] Springtime light in deciduous forests nurtures a panoply of wildflowers known as spring ephemerals—onegrass (*Anemone nikoensis*) is typical—as well as many kinds of herbs and bamboo grass (*Sasa*) throughout the summer. By one estimate, two thirds of plant species considered native to Japan are found in cool temperate forest floors, especially beneath beeches.[44] In cleared areas, dryfield farming of grain and forage crops predominates in this climate.

Warm temperate zones along the Pacific coast of Honshu, Shikoku, and Kyushu feature serried ranges of cryptomeria (often mistaken for cedars), many in perfect rows via silviculture techniques dating to the Edo period and more recently renewed to replenish stocks felled during World War Two. Camphors (*Cinnamomum camphora*), broadleaf evergreen oaks (*Quercus acuta*), and chinquapins (*Castanopsis cuspidata*) flourish in dark, damp forests, with various orchids, rhizomatous fringed iris (*Iris japonica*), and species of the hollyhock genus (*Alcea*), sometimes considered "grotesque and a bit unpleasant,"[45] on the forest floor. Red pines (*Pinus densiflora*) and black pines (*Pinus thunbergii*) became well established in warm temperate areas and have been the predominant pines for landscape gardening, although now they are imperiled by predators.[46] Various subspecies of cherries (*Prunus serrulata*) grow well in this climate; the famous Yoshino cherries (*Prunus x yedoensis*) are thought to be clones

Figure I.3 Black pines planted in 1940 in the Palace Outer Garden, Tokyo, 2009.

of a hybrid of unknown origins near the main gate of Tokyo's Ueno zoo.[47] Many warm-temperate trees are linked with counterparts in southern China and Southeast Asia. Ginkgos (*Ginkgo biloba*), for example, originated in China 270 million years ago and, because they are pollution-resistant, are now the most commonly planted in Japan's cities, comprising 11 percent of all street trees.[48] Varieties of azaleas (*tsutsuji—Rhododendron dilatatum* or *R. japonicum* and *satsuki—Rhododendron lateritium*) and other species of the *Rhododendron* genus are common in warm temperate regions, as are meadows of bush clover (*Lespedeza bicolor*) and cultivated angiosperms such as chrysanthemums, lilies, and irises. Wet-rice agriculture thrives in the warm temperate regions of Japan, as do dandelions, crabgrass, and other volunteers considered weeds.[49]

Southern Kyushu and Okinawa form a subtropical area of evergreen oaks, chinquapins, cycad palms (Cycadophyta), fern palms (*Cycas revoluta*), fruit trees, camellias (*tsubaki*), laurel (*Aucuba japonica*), and colorful tropical flowers.[50] Mandarin oranges (*Citrus tachibana*) are cultivated from western Honshu, Shikoku, and Kyushu to Okinawa. Pyracanths (*Pyracantha angustiflora*) with bright red or orange berries grow rampantly in Okinawa. Mitrasteman (*Mitrasteman yamamotoi*) is a broadly distributed parasite with a white bulbous shape that feeds on beeches, oaks, and *itajii* chinquapins (*Castanopsis sieboldii*) in subtropical districts. Many of Japan's subtropical plants were borne as seeds by ocean currents from southern China or Southeast Asia.[51]

Viewed from top to bottom or side to side, every region of Japan incorporates vegetation that evolved in response to vast ecological changes in temperature, rainfall, climate, soil conditions, and animal behaviors across the geologic ages. The porous borders between climate zones have shifted up and down mountainsides as well as north and south depending on warming or cooling trends across geologic time, with major consequences for the distribution of both plant and animal species. Yet the Japanese islands have been able to host, indigenize, and introduce variants of thousands of plant species from continental Asia and elsewhere while also evolving varieties that are distinctly local, endemic to Japan, and widely considered "native."

Natives and Invasives

Plants are social beings that flock together, avoid competing for nutrients with other plants of the same species, and communicate chemically to defend against insects or pathogens. Most plant biologists today focus on cells and genes, but bioecologists and plant ethologists prefer to examine how flora behave in specific environments. Plants respond to light and darkness, touch, sounds, and other sensory stimuli, endowing them with powers of sustainability and regeneration.[52] For example, a number of plant species on Itsukushima (Miyajima) in Hiroshima bay, such as Japanese andromeda (*Pieris japonica*) and Japanese star anise (*Illicium anisatum* or *I. religiosum*), defend against the island's deer population via thorns, poisons, and foul odors.[53]

Although nominally sessile (rooted in place), across deep time plant species are considerably mobile; as early as 1880 Charles Darwin (1809–1882) published *The Power of Movement in Plants,* a work still in print today that has inspired numerous studies of plant mobility.[54] Camille Parmesan and Gary Yohe have analyzed 1,700 plant and animal species from around the world, finding that their ranges have shifted nearly 6.5 kilometers per decade toward the poles because of climate changes.[55] Other research suggests that certain plant species could shift tens or hundreds of kilometers within 50 to 100 years, although migration is usually as slow as five kilometers per century.[56] Long-range movement of floral species has occurred for many millions of years through dispersal of seeds by winds, by ocean currents, and by birds and other animals, and more recently through coincidental or deliberate human agency. Japan is not alone as a land of plants in motion, but it is one of the most prominent because of the great favor plants from that country continue to enjoy around the world.

Of Japan's approximately 7,500 species of wild-growing plants, about 1,860 seed-bearing species, ferns, and mosses are thought to exist naturally nowhere else and are considered natives.[57] Some fraction of these must have evolved indigenously from continental ancestors into distinct species; only those introduced by people are considered nonnatives by ecologists. Landscape gardens in Japan have a strong preference for native plants, except for chrysanthemums, which came from China in the sixth century CE. The majority of native flora are found in the cool temperate beech forests, such as katsuras (*Circidiphyllum japonicum*) and Japanese bigleaf magnolias (*Magnolia obovata*). Others exist on islands such as Chichijima and Hahajima in the Ogasawara group, on Yakushima, and at sites in Okinawa.[58] Many native species are naturally occurring hybrids and are more widely found within the country than nonhybrids; species with narrow distributions run a considerable risk of extinction.[59] Notably, experts at the Tsukuba Botanical Garden estimate that two thirds of orchid taxa native to Japan are currently threatened.[60] Together with naturalized plants now considered indigenous, these native flora form a prolific endowment of plants unsurpassed anywhere else in warm temperate climates.

One result of this abundance is that Japanese species, such as pachysandra (*Pachysandra terminalis*), are often preferred in North American gardens, not local counterparts such as the Allegheny pachysandra (*Pachysandra procumbens*). Botanist David Fairchild (1869–1954) pointed out in 1903 that Japanese horticulturalists since the mid-Edo period emphasized selection and improvement, whereas American specialists did little to improve native species in the nineteenth century.[61] This helps explain why Japanese plants came to enjoy such popularity in the United States. Among flowering shrubs the Hirado azalea (*Rhododendron x pulchrum*) from Kyushu is especially favored by overseas horticulturalists, particularly in the United States, where by one count more "Japanese" gardens were built in the early twentieth century than in Japan. Most were products of Western landscape architecture and seemed Japanese in "only the most perfunctory ways"[62]—an outcome consistent with the spirit of gardens in Japan to build with local materials.[63]

The movement of plants from one land to another often stirs debate over the concept of invasive species, partly because a plant can leave natural predators behind and encounter few if any restraints after it spreads to the new location. The debate occurs in two registers, geological timescales (tens of thousands of years or more) and ecological timescales (decades to hundreds of years). The environmental writer Elizabeth Kolbert points out that plants and animals migrated globally alongside human ancestors during Pleistocene times (2.58 million–12,000 years ago); at present perhaps 10,000 species are moving around the world every day as ballast water—a kind of "mass invasion event"[64] from the perspective of ecological time. The U.S. Office of Technology Assessment defines an invasive as "an alien species whose introduction does or is likely to cause economic or environmental harm or harm to human health." The United States has roughly 17,000 native plant species and 5,000 to 6,600 nonnative species; some but not all of the latter are considered invasive.[65] The biologist Yi Li and associates estimate that 50 percent of all invasive plants and 85 percent of invasive woody plants arriving in the United States over the past 150 years were originally imported for landscape beauty; most of the rest were introduced for agricultural purposes, animal forage, or erosion control.[66] Many have subsequently escaped managed cultivation in gardens, parks, or farmlands to become naturalized. Yet inasmuch as ornamental horticulture is the fastest-growing sector of American agriculture, "completely prohibiting new imports and totally banning the use of all invasive and potentially invasive ornamental plants are socially, politically, and economically unfeasible."[67] Instead, the United States spends tens of billions of dollars annually to eradicate or contain species considered harmful to indigenous plant life.

Yet the distinction between native and nonnative species in North America is fuzzy, historically shortsighted, and to a degree culture-centric. Native plants are supposedly those present before the first European contact, even though most were themselves invaders as they moved into North America at some point in deep geological time. One town asserts that "invasive plants are plants *not native* to the New England ecosystem. . . . Native plants are the foundation of our ecosystem, supporting insects, birds, and other wildlife."[68] In the San Francisco region, aficionados of native plants "yearn to restore the treeless 'native' grassland that greeted the first European settlers of the Bay Area in 1769,"[69] regions of northern California where large grazing animals maintained grasslands for hundreds of thousands of years before indigenous Native Americans began clearing trees for better hunting well in advance of the first European settlement. Biologist Mark Davis calls the opposition to invasives "ecological fundamentalism, the notion that the purity of the past has been polluted by outsiders."[70] Historian of science Stephen Jay Gould (1941–2002) scoffed at the fear of invasive species as "romantic drivel,"[71] while botanist Peter Del Tredici points out that "re-creating a lost landscape is an impossibility, because the conditions under which these landscapes evolved no longer exist. The world is a totally different place as a result of human activity. There's no going back in time."[72] Another specialist, David Boufford, agrees that the term *invasive plants* is "not quite appropriate" and may refer only to certain vegetation at the edges of forests.[73] Yet a mainstream ecological outlook holds that indigenous plants and animals found nowhere else need protection against invasives through restoration of local ecosystems. Such protections aim to offset the deleterious impact of invasive species on biodiversity and ecological stability, although some native wildflowers and herbs grown in nurseries and then transplanted in California restoration sites have been found to be infected with pathogens resistant to fungicides.[74]

Plants from Japan that the International Union for Conservation of Nature considers invasives abroad include kudzu or Japanese arrowroot (*Pueraria montana* var. *lobata*), Japanese knotweed (*Fallopia japonica* or *Reynoutria japonica,* previously *Polygonum cuspidatum*), and wakame seaweed (*Undaria pinnatifida*). Spread by spores in marine ballast, wakame thrives in areas lacking sea urchin (its predator) and clogs fish farming facilities.[75] Other invasives in parts or all of the United States are Japanese barberry (*Berberis thunbergii*), multiflora rose (*Rosa multiflora*), winged euonymus or burning bush (*Euonymus alatus*), Japanese privet (*Ligustrum japonicum*), and dayflower (*Commelina communis*). Important in Japan but sometimes regarded as invasive in the United States are the camphor tree (*Cinnamomum camphora*), spiraea (*Spiraea japonica*), Japanese honeysuckle (*Lonicera japonica, Lonicera japonica* var. *halliana*), and Japanese wisteria (*Wisteria floribunda*), the latter of which arrived in the United States as early as the 1830s and has long since been naturalized.[76] In short, the great majority of introduced Japanese garden plants pose no threat to North American ecosystems, but those few that prove menacing can cause extensive damage.

Kudzu is a woody vine seldom cultivated in Japan because it is so vigorous. It was introduced to the United States at the 1876 centennial in Philadelphia and widely promoted by federal agencies to cut soil erosion and as cattle fodder until after World War Two. Its vines can climb twenty meters in a season, girdling and finally overtopping shade trees, weighing down canopies to the point of breakage. In 1953 the U.S. Department of Agriculture stopped promoting kudzu as a cover plant, and Congress listed it as a federal noxious weed in 1997.[77] Japanese knotweed is a large plant 3 to 4 meters tall with bamboo-like hollow canes and heart-shaped leaves used in herbal medicine that thrives in volcanic lava and ash but has many predators in Japan. It was introduced to Europe from Nagasaki by Philipp Franz von Siebold and made its way to England in 1825 and to southern Poland by 1882. It arrived in the United States from Britain as a garden plant to stabilize soils and was naturalized by the 1880s. Frederick Law Olmsted (1822–1903) planted Japanese knotweed in Brooklyn's Prospect Park and the Emerald Necklace in Boston. This plant deprives other species of sunlight and spreads via rhizome to break through soil and paving, sometimes even damaging buildings. Great efforts to control its spread have been underway since the 1990s, including releasing Japanese jumping plant lice (*Psyllidae*) in Britain starting in 2010 to attack the leaves. In Boston volunteers have been trying to eradicate huge Japanese knotweed plants since 2007.[78] Seen from the perspective of deep history, such human interventions in the global movement of plants across the ages may seem quixotic at best and quite possibly futile. But from the standpoint of ecological time, the few species that became invasive pose potentially serious threats to biodiversity and local primary industries, the more so because climate change outpaces the capacity of plants to evolve adaptations, leading to high rates of extinction.

At the same time that certain plants from Japan are regarded as invasive in other countries, Japan itself lists a number of species introduced from abroad as dangerous imports. Yet Ōba Hideaki, the dean of Japan's botanical historians, is quick to point out that there is no good or evil among plants; all have the same value in natural history.[79] Ōba and other Japanese biologists are careful to distinguish among native, introduced, naturalized, and invasive species, mindful that the boundaries among them are blurry. Native (*zairai*) plants are those found in Japan before 1800, many of which arrived from abroad in earlier epochs. Especially numerous are species of chrysanthemum and rice plant genera, both arriving at the dawn of Japanese history. Introduced (*gairaishu* or *inyūshu*) species are those arriving during the past two centuries, deliberately or accidentally through human activity, taking root from random seeds of flowering plants scattered by winds, from grain seeds imported on a huge scale, or from seeds piggybacking on beans from abroad.[80] When they survive and begin to grow in the wild, they are considered naturalized (*kika*). A small fraction of Japan's 2,000 naturalized plant species end up causing ecological disruption and are called invasive (*shinryakushu*) by government authorities and many farmers and horticulturalists.[81]

Even today plants naturalized during Japan's prehistory can harm forests or farm crops. A huge invasive from abroad is the tortoiseshell bamboo (*Phyllostachys heterocycla* or *P. edulis*), a giant timber bamboo imported to Hakodate from China in the early eighteenth century in order to cultivate its edible (*edulis*) bamboo shoots. It quickly started to grow wild, with rhizomes extending more than twenty meters long and canes reaching twenty-eight meters high. Grasses brought to Japan for pasturage in the mid-nineteenth century soon grew wild beyond their original cultivation areas and crowded out existing species.[82] Especially noxious is *Spartina alterniflora* (smooth cordgrass or Atlantic cordgrass), an herbaceous species that may have arrived with ballast water or spread from coastal China via floating seeds. In 1972 the Osaka media had a field day publicizing the "dandelion wars" when two species of the dandelion genus (*Taraxacum*), recently arrived from Europe, wiped out existing species and spread aggressively from countryside to city.[83]

The numbers of plant species from overseas introduced to Japan soared after World War Two, and by now most plants growing wild in cities, such as goldenrod (*Solidago spp.*), are relatively recent foreign arrivals. In 2002 the Ecological Society of Japan (Nihon Seitai Gakkai) published a list of "Japan's Hundred Worst Invasive Species from Abroad."[84] Two years later the Diet enacted the Invasive Alien Species Law (Gairai Seibutsuhō), effective June 2005, that enumerated invasive animals and listed many foreign plants as enemies of nature threatening the Japanese ecosystem, especially agriculture and landscape gardening. The law called for no importation, no release, and no propagation of invasives (*shinryakuteki gairai shokubutsu*), with emphasis on twelve particularly harmful plants, including species of the dandelion and chrysanthemum genera.[85] Plants in Japan offer little resistance to predators from abroad, such as the pinewood nematode (*Bursaphelenchus xylophilus*), which entered Nagasaki from North America in the early twentieth century and has destroyed a majority of the country's red and black pines.[86] Living in an island country shielded for many millennia from continuous contact with overseas ecosystems, Japanese flora has been especially vulnerable to displacement by or hybridization with newly arriving plant species, giving special acuteness to the understanding of *invasives* in that land today.

Reconnoitering the history of Japan's engagement with its rich botanical inheritance and the movement of its plants globally is complicated by distance, both chronological and geographical, by uneven documentation, and by the low priority accorded the history of science for much of the period treated here. *Land of Plants in Motion* relies partly on paleo- and ethnobotanical research reports, historical maps, illustrated flora by Japanese brush and photo artists, primary documents by European and American plant collectors, nursery catalogs, library and herbaria holdings, plant inventories, and profuse statistical data. It also treats as texts Japan's many botanical gardens, arboretums, greenhouses, natural parks, fields, forests, and mountains in order to offer historical perspective on the changes across time examined here. For information and insights about the plant sciences, I've been guided by the scholarship and recollections of Japanese academic specialists on the topic, supplemented by my own rambles around many of Japan's botanical gardens and arboretums during research visits to that country. No single methodological approach or theoretical position can adequately interpret the multiple meanings of plant life across time in Japan. Instead the reader can be confident, as Victor Brombert writes, that "eschewing a dogmatic approach and stressing diversity and variation do not preclude a search for underlying patterns and common tendencies."[87] These patterns and tendencies form the matrix for the chapters that follow.

One

East Asia's Plants in Geological Time

Trees, shrubs, herbs, flowers, and other components of Japan's flora today are ubiquitous, lush, and usually healthy—and survivors: they and their direct ancestors across hundreds of millions of years have overcome fires, floods, predators, earthquakes, volcanic eruptions, tsunamis, tectonic shifts, planetary warming and icing, and five vast extinction events, the most catastrophic of which, "the Great Dying" or Great Extinction, obliterated 96 percent of living species about 252 million years ago. The earth's current flora and fauna are survivors of a long and highly selective evolutionary process; today's global biota is almost unrecognizably altered from what life looked like 10 or 100 million years ago. In a number of cases, terrestrial plant species that once lived throughout the northern hemisphere later became extinct everywhere except East Asia, occasionally surviving only in what became Japan. Two notable species that anciently achieved wide distribution have survived today as Lazarus species (what Charles Darwin called "living fossils"), the ginkgo and the metasequoia.[1] Evidence of earliest ginkgo-like trees dates to 270 million years ago in South Africa, and as recently as 50 million years ago various forms flourished globally, but ginkgos later retreated to southern China, and today only a single species, the maidenhair tree (*Ginkgo biloba*), survives—as a popular boulevard tree in cities around the world. The deciduous conifer metasequoia (*Metasequoia glyptostroboides*), which dates to at least 150 million years ago, at one time thrived in what is now Britain and many other parts of the globe, then was driven by glaciers into a remote refuge in southwestern China. After its discovery in 1946, metasequoia seeds have been distributed throughout the world to cultivate this eighty-meter-high member of the redwood subfamily.

This chapter tells the fascinating story of how plant life evolved in the East Asian floristic zone from earliest times to the historic period, highlighting the ancestors of species that survived to flourish in modern Japan. Then it examines the striking similarities at the genus level between many East Asian plants and those in eastern North America, although individual species have often evolved quite differently from their counterparts across the Pacific.

Paleoceanographers, paleoclimatologists, paleogeologists, and paleobiologists study sediment cores, rock formations, and the fossils and pollens contained within them in order to understand past environments, including the evolution of plants, animals, and other organisms ever since the earliest forms of life appeared in the sea nearly 4 billion years ago. One recent exemplar is Komiya Tsuyoshi, a professor of earth sciences at the University of Tokyo, who in 2017 led a team that discovered grains of graphite and other types of carbon in rock fossils on Labrador dating to 3.95 billion years ago, yielding isotopes that proved to be "biogenic." Komiya announced, "What we found was the existence of life. We haven't yet pinned down the type of organism. That's the next task for us to tackle."[2] The first animal footprints on earth may date to 530 million years ago, followed by terrestrial flora about 485 million years ago, when rudimentary plants, all descended from a single ancestor, moved from water onto a barren landscape pelted by harmful ultraviolet rays, an inauspicious start to a long evolution from freshwater algae into mosses, ferns, and vascular plants with seeds, flowers, and new organs.[3] A few of their remote descendants survived uncounted perils to thrive in what became the Japanese islands and also took root in families and genera in East Asia and throughout the world.

The Slow Progression of Plants in Japan

Cooksonia, an extinct genus of slender ground cover, was the first vascular plant to live on land, thriving in equatorial districts from 433 to 393 million years ago. *Rhynia,* a single-species genus of fernlike vascular organisms that lived about 400 million years ago, is considered one of the earliest land plants with a descendant in Japan today, the whisk fern (*Psilotum nudum*), an evergreen widely distributed in the Edo period.[4] Both gymnosperms (seeded plants without ovaries, such as conifers, ginkgos, and cycads) and angiosperms (flowering plants with seeds within female reproductive organs, including most trees) evolved from ferns and remain robust today. The first trees date to about 380 million years ago and the earliest seeds to about 360 million years ago.[5] The blackish-green scouringrush horsetail (*Equisetum hyemale*), a rough-textured, wetland herbaceous plant that originated in the verdant, insect-rich Carboniferous period (359–299 million years ago), was once the most widely distributed land plant on earth and survives today throughout temperate zones in Eurasia and North America, often used as sandpaper. A related but structurally dissimilar genus, now extinct, was *Calamites,* a tree in the shape of an inverted bamboo broom that grew more than 30 meters high in the moisture and great global heat of 300 million years ago, a time when sharks patrolled the oceans.[6] Various scaly trees, grasslike herbs, ferns, and mosses from the same era have escaped extinction and are common today.

Ginkgos, dating to the Permian period (299–252 million years ago), look quite different from cycads but are similar in structure and reproduction. Cycads grow on seaside cliffs and are often called fern palms but actually are neither ferns nor palms; instead they form a distinct group of gymnosperms with woody trunks topped by splayed leaves, emerging 280 million years ago, 10 million years before ginkgos. Permian ginkgos resembled conifers, and by the Jurassic period (201–145 million years ago) a range of ginkgo-like forms existed across the northern and southern hemispheres. The sole surviving species, the maidenhair tree, shows very little evolutionary change from its Jurassic forebears.[7] Leaf fossils reveal that ginkgos were still abundant at the start of the Tertiary period (66–2.6 million years ago) and flourished in the Rocky mountains, Oregon, and near the north pole during a stretch of great planetary warmth about 50 million years ago. Then global cooling and drying, which began 35 million years before the present, eventually drove ginkgos from Europe and North America by 15 million years ago, and within the next 10 million years they retreated to East Asia along with metasequoias, katsuras (*Cercidiphyllum japonicum*), and other trees.[8] Likewise the *Sequoia* genus now found only along the North American Pacific coast was widely distributed in the northern hemisphere as late as the Pliocene epoch (5–2.6 million years ago).[9]

Figure 1.1 The Meiji Shrine Outer Garden, dating to 1926, features four rows of ginkgos leading to the Meiji Memorial Picture Gallery, 2009. Ginkgos are the official tree of Tokyo and a favored boulevard planting.

Ginkgos have long been valued in East Asia for their fruits (Jp. *ginnan*), oils, and medicinal efficacy. Their resistance to heat, cold, smog, and vehicular crashes is prized in cities around the world, to which the ginkgo was reintroduced by plant explorers starting in the eighteenth century, at first as "a horticultural novelty."[10] Six ginkgos near the epicenter of the Hiroshima atomic bombing in August 1945 demonstrated their resilience by surviving the blast with little damage because of their deep roots, and soon they began to bud.[11] Equally famous is the thousand-year-old ginkgo at Kitakanegasawa, Fukauramachi, in Japan's Tohoku region, rated Japan's second-largest tree with 22 meters' girth at chest level and a height of 31.8 meters.[12]

Ginkgos and a small fraction of other hardy animal and plant species have managed to survive cataclysms throughout geologic time. Apart from the Great Oxygenation Event about 2.4 billion years ago, which introduced oxygen to the atmosphere and killed off obligate anaerobic organisms, five huge mass extinctions between 440 and 66 million years ago wiped out a majority of flora and fauna each time. Fossils suggest that the few surviving species needed hundreds of thousands of years to recover and diversify because the global ecosystem took that long to rebalance itself after such catastrophic damage.[13] The most devastating was the Great Extinction of 252 million years ago, probably triggered by carbon dioxide released from volcanoes in Siberia that produced giant lava floods and caused the marine ecosystem near what is now Japan to collapse. Scholars estimate that marine life reestablished itself near Japan within 5 million years, but elsewhere recovery took at least twice this long.[14] Most well-known is the Cretaceous-Triassic Extinction 66 million years ago, which eliminated dinosaurs and was caused by an asteroid impact on the Yucatán peninsula and nearly simultaneous volcanic eruptions. At least fifteen lesser extinction events have also occurred since life began on earth.[15]

As vascular plants evolved about 400 million years ago and seed plants arose 40 million years later, large ferns developed like trees, and smaller ferns continued to thrive. Seed plants and ferns have the same basic structure—roots, stalk, leaves—but seeds allow a species to propagate broadly, even in arid environments.[16] Among gymnosperms, the first conifers lived more than 300 million years ago, but the now extinct Voltziales, one of the few plant orders to survive the mass extinction event of 252 years ago, is the source from which subsequent conifers evolved. Botanists consider the Japanese umbrella pine (*Sciadopitys verticillata*) especially important for the evolution of gymnosperms. This species thrived in Europe and Greenland from 250 million years ago until the extinction event 66 million years ago, then, faced with glaciation and continental separation, it eventually retreated to Japan, where it grows at higher elevations amid forests of beeches and oaks.[17] Cryptomeria, another conifer, has left few fossils but appears to date to more than 200 million years ago. Northern conifer forests with sequoias, metasequoias, firs, and eventually pines established themselves worldwide during cooler ages nearly 220 million years before now. Without the sequoias and metasequoias but with abundant Japanese spruces, these forests remain prevalent in Japan. Biologists consider the *Ephedra* genus (Jp. *maō*) to be the most highly evolved among all the gymnosperms. These nonconiferous plants yield ephedrine, a stimulant used during anesthesia and sometimes to treat asthma.[18]

By 100 million years before now, fossil evidence confirms, even Antarctica was dotted with sparse forests,[19] and many types of plants now common in Japan were well established around the planet. Because of global warming, deciduous conifers and other deciduous species were found only near the poles. Most trees were shorter than their descendants today, providing forage for plant-eating large dinosaurs in polar woodlands and smaller dinosaurs in dense forests at lower latitudes.[20] Species of the oak genus such as Konara oaks (*Quercus glandulifera*) and Mongolian oaks (*Quercus mongolica*) trace to relatives in the Cretaceous period (145–66 million years ago), as do beech and chestnut genera. Many birch fossils date to this period, as is true of willow and poplar genera. Katsura trees originally existed as two species, one of which is now lost; these deciduous broadleaf trees emerged at the beginning of the Cretaceous period and thrived throughout the cooler regions of the northern hemisphere until 2.6 million years ago, after which they were found only in China and Japan. Species of the tulip tree genus *Liriodendron* similarly existed throughout the northern hemisphere in Cretaceous times, then began to wane until being confined today to East Asia and the North American east coast. Herbaceous (nonwoody) plants with simple flowers may have abounded on the dim floors of Cretaceous forests, but few fossils exist to confirm their characteristics.[21]

The extinction event of 66 million years ago devastated insects and other animals at sea and on the land, leaving crocodiles as one of the few surviving vertebrate species. Although historian of botany Peter Crane believes the effects on plant life in the long run "were seemingly much less profound,"[22] biologist Robert Askins points out that at a large site centering on Montana in continental North America, 80 percent of plant species became extinct, causing the deaths of plant-eating insects as revealed in surviving flora: "The frequency of insect damage on leaves remained low for the next million years."[23] Deciduous trees subsequently fared better than broadleaf evergreens, as did forests at high latitudes and woody plants in wetlands. In the Tertiary period deciduous hardwoods became dominant at midlatitudes in much of eastern North America as well as in upland cool temperate areas in Japan, while evergreen oaks and chinquapins flourished in warmer temperate parts of western Japan.[24]

Angiosperms developed slowly from gymnosperms and eventually surpassed them in number of species by a huge margin, so that today "they comprise the autotrophic foundation on which almost terrestrial ecosystems are built"[25]—and thus form the basis of human survival. Their origins can be dated by pollen grains at sites in Colorado, Greenland, South America, and Africa to at least 135 million years ago. Some attracted insects and birds to spread pollen; others were pollinated via wind or water, allowing flowering plants to colonize and diversify widely, eventually displacing conifers and other gymnosperms as the most numerous and highly evolved plant forms. By the time of human settlement, by one estimate, flowering plants accounted for 90 percent of Japan's terrestrial plant species, gymnosperms 8 percent, and ferns 2 percent.[26] Japanese researchers have studied angiosperm fossils in Hokkaido for more than a century, using them as microcosms of plant life that girdled the globe 100 million years ago; since 1980 new molecular techniques and computer-assisted data analyses have speeded studies of angiosperm fossils from around the world.[27]

Flowering plants can be considered monocots (sprouting single leaves) or eudicots (sprouting multiple leaves), although this distinction has grown decreasingly useful to researchers as the field of systematics has progressed since the 1990s. Monocots account for one fifth of angiosperms, and eudicots for almost all the rest. The lily genus (*Lilium*) is an ancient monocot from the Cretaceous period, most species of which were herbaceous and left almost no fossils, but the closely related genus *Smilax* developed woody stalks and is widely found in the fossil record from that time. Nipa palms (*Nypa fruticans*) are another early monocot that was found near mangroves in tidal estuaries around the world 100 million years ago but today is limited to tropical areas from Okinawa south.[28] The peony genus (*Paeonia*) ranks among the first angiosperms to evolve in the Cretaceous period and today counts about forty species throughout the northern hemisphere. The herbaceous hasu lotus family (*Nelumbonaceae*), a key presence in Buddhist teachings, is the oldest form of water lily and grew in the Cretaceous period in Europe, North America, and East Asia. Today only two species remain, one of which (*Nelumbo nucifera*) is widely distributed in Australia, China, and southeastern Europe as well as Japan.[29] Among angiosperm trees, the beech and walnut families, both with simple flowers, trace to the Cretaceous period, as does the witch hazel genus (*Hamamelis*). Pollen samples of the magnolia genus have been dated to the Jurassic period (201–145 million years ago), whereas in Japan evidence of magnolias is sparse until the Tertiary period. Widely admired as ornamentals, these handsome trees thrive today in East Asia and the Americas.[30]

Japanese botanists sometimes refer to the age of huge volcanic eruptions 90 to 80 million years before now as the Hiroshima change, a time of much magma flow, white sands, and the formation of granite along the northeast coast of the Asian continent. These events occurred as angiosperms were growing dominant among flora, both globally and in what later became Japan. Pines also likely were found along what eventually formed as Japan's inland sea.[31] Askins concludes, "70 million years ago the earth was just as geographically diverse as it is today, with sand dune deserts, tropical forests, and swamps, each with a distinct assortment of dinosaurs and other vertebrates."[32] This diversity must have been just as true of Japan and the Asian continent to which it was attached at the time.

After terrestrial vegetation recovered from the mass extinction of 66 million years ago, many of the world's modern flowering species emerged. Japan is an especially fruitful site for understanding their development during the next 15 million years, one authority noting, "There is probably no area of its size in the world which has as many fossil plants of Tertiary age as Japan, or in which these fossils show finer preservation."[33] Volcanic sediment, extensive coal deposits in Hokkaido and Kyushu, and climates favorable for forests helped preserve fossils and have drawn the world's paleobotanists to Japan in recent times. Between 56 and 35 million years ago the planet warmed by 5 to 10 degrees centigrade, inviting broadleaf evergreens to advance from the tropics northward—remarkably, there is even evidence that Hokkaido featured bananas and palm trees.[34] During those warm years East Asian floras were less diverse than today, and many species may have been identical with those in Europe and North America.

By 30 million years ago a number of genera became differentiated, closely resembling those in present-day Japan.[35] Both camellia and linden genera were widely distributed in the temperate northern hemisphere, yet today the greatest species diversity of each is found in East Asia. The plane tree genus (*Platanus*) dates to the Cretaceous period, and its relatively few species were well represented globally by the mid-Tertiary period, surviving today in Japan as the large-scale plane or sycamore (*Platanus racemosa*) and as the cultivated London plane (*Platanus x acerifolia, Platanus x hispanica*), a familiar high-maintenance tree lining boulevards and plazas in Japan since the late nineteenth century. Elm and zelkova genera, both part of the *Ulmaceae* family, left behind fossils from the Tertiary period in Japan and throughout the temperate northern hemisphere, ancestors of the various species of elms and zelkovas today.[36] The same is true of huge numbers of woody and herbaceous angiosperms from the Tertiary period with descendants in modern Japan.

Herbaceous plants probably evolved from woody ancestors as climate changes took hold. Various species seem to have emerged during the Cretaceous period but cannot be properly documented because fossil evidence is lacking. In fact, by one measure, 92 percent of genera currently alive in East Asia lack fossil records from any period, mainly fragile herbaceous plants that lack structure susceptible to preservation. Three fifths of seed plant genera considered endemic to East Asia today are herbs, yet almost none appear in fossils; in contrast, about one fifth of woody plants are known from fossil records. Unlike most gymnosperms and other woody plants, many herbs seem to have originated in East Asia, and some never spread far afield.[37] They probably included groups from dank forest floors, disturbed forest sites, other stressed environments, and freshwater ponds and streams. Throughout the Tertiary period, new plant species kept taking hold while others kept disappearing.[38]

About 35 million years ago the warming trend peaked, and the world began a long, slow swing back to cooler temperatures, losing 10 degrees centigrade by 2 million years ago, probably because of continental drifts, the formation of new mountain ranges, changing ocean currents, and reduced carbon dioxide.[39] Japan experienced a dramatic geological reordering starting about 20 million years ago, sometimes called the green tuff (volcanic rock) changes. Amid violent volcanic eruptions, some stretches of land subsided into the ocean and others heaved up, gradually broadening the Sea of Japan, creating inland lakes, and shaping the Japanese islands of today, although western Japan was still attached to the continent.[40] An undersea forest discovered in 2015 off Aomori prefecture yielded microbes within a seam of coal 3.6 kilometers below sea level, evidence that woodlands from southern Hokkaido to northern Honshu sank with movements of the Pacific plate.[41] Fossils from the green tuff upheavals confirm that the cooling underway at the time caused broadleaf evergreens to retreat southward and temperate genera to predominate, much as they do in the northern hemisphere at present.[42] Grasses and flowering plants prospered, but many species were lost as temperatures fell. Nonetheless 76 percent of plants found in fossils from 3 million years ago live in Japan today, with evergreen oaks especially prominent.[43] While the cooldown continued, many plants from East Asia spread into Europe and North America, and some flowed the other direction. Other genera, such as katsuras and metasequoias, abandoned a global presence and retreated to East Asia.[44] By 8 million years ago, Askins believes, the cool temperate forests of Japan resembled those of Japan and eastern North America today, populated by maples, beeches, ashes, elms, zelkovas, and conifers.[45] The interchange between East Asia and eastern North America may have ended about 5 million years ago, after which speciation underwent different rates and trends of separation in the two regions, even though "in many cases Japanese species . . . have the same chromosome number as those of the corresponding taxa in Europe or North America."[46]

Flora and fauna in Japan were not devastated by glaciers during the most recent glacial period, known as the Quaternary or Pleistocene glaciation (from

2.6 million to 11,700 years ago), but they were affected by the intermittent warming and cooling during this period that had much greater impacts elsewhere in the northern hemisphere. Geologists currently recognize seven ice ages, five of them considered major, since the earth was formed. The Huronian (2.4–2.1 billion years ago) immediately followed the Great Oxygenation Event, when oxygen combined with preexisting methane to form carbon dioxide and water, cooling the planet and causing mass extinction. The largest ice age, known as Cryogenian (720–625 million years ago), may have covered all or nearly all of earth. The Andean-Saharan (450–420 million years ago) and Karoo (360–260 million years ago) ice ages both took place before supercontinents broke up about 175 million years ago and their components began to assume separate positions. The Current ice age (34 million years ago to the present) is so-called because Antarctica has been continuously covered by an ice sheet; the Quaternary glaciation comprises the last 8 percent of the Current ice age, when the formation of ice at its maximum caused sea levels to fall 120 meters.[47] Eight glacial cycles during the past 740,000 years spread ice sheets up to four kilometers thick across North America, Siberia, and Europe. The most recent cycle lasted from 110,000 to 11,700 years ago but within Japan created glaciers only on the highest mountains in Hokkaido, which was attached to Sakhalin and the continent until 20,000 years ago, and on peaks in the single Honshu-Shikoku-Kyushu landmass.[48] The limited damage caused by Quaternary ice cycles, one authority believes, "is largely responsible for the richness of Japanese flora and the diversity of the vegetation" since.[49]

During ice cycles plants and animals seeking refuge crossed into Kyushu from Korea and into Hokkaido from Sakhalin, but the depth of the Tsugaru strait made it difficult for species to transit between Hokkaido and Honshu. As a result, somewhat different plant species developed on mountain slopes in Hokkaido compared with those on the hillsides of the Japan Alps. In the Honshu-Shikoku-Kyushu landmass some alpine plants moved south to flee the cold, then north to the cool mountains during interglacial periods.[50] Almost all East Asian tree genera survived the ice cycles; fossils reveal that the Japanese larch, whose range normally extended to Siberia, moved south during ice cycles to what today is warm temperate Nishinomiya in western Honshu but at that time was cold and dry. At the height of the last global ice cycle about 20,000 years ago, Japan was replete with various species of fir, spruce, hemlock, pine, and birch genera. The mountain avens (*Dryas octopetala*), an evergreen flowering shrub found in arctic areas such as Alaska, spread into Japan during ice cycles and survives today in the coldest mountains.[51] The narcissus anemone (*Anemone narcissiflora*) is another alpine angiosperm forced south in Japan and left behind by glaciers. But many other plants may have moved south independently of ice cycles because the glaciation was relatively benign. Certain alpine species likely crept south into barren spaces created by volcanic eruptions.[52] In short, the times, if not always the ice itself, were key for the evolution of Japan's current floral species: some died out, many moved, some developed in differentiated ways by adapting to new local conditions.[53] Overall far more East Asian plant species persisted during the Quaternary glaciation compared with those in Europe and North America, yet quite apart from plant collecting and commerce by Westerners in recent centuries, the floras of Europe, East Asia, and eastern North America continue to represent a number of the same genera as they have for tens or hundreds of million years.

As the most recent ice cycle began to recede 15,000 years ago, Japanese oak, birch, maple, and other deciduous hardwood genera gradually moved north and diversified. At the same time, beech species that were once widely distributed in the northern hemisphere before the Quaternary ice cycles and then retreated to East Asia now became prominent in Japan, especially Japanese beech (*Fagus crenata*) and Japanese blue beech (*Fagus japonica*).[54] As one study puts it, "The paleobotanical literature abounds with reports of genera identified from the Tertiary of Europe and North America that are now living only in Asia."[55] Cryptomeria and broadleaf evergreens propagated widely in the Kanto area near Tokyo about 7,000 years ago in response to mild ocean currents in the Pacific and Sea of Japan, averaging 2 degrees centigrade warmer than today.[56]

Grasslands abounded in much of Japan under the cool, dry conditions of the last glaciation; thereafter Jōmon-era (12,000–1000 BCE) peoples in Kyushu fabricated thatch, one of the earliest instances of the current practice called *satoyama,* using quasiforested woodlands for firewood, charcoal, mulch, fertilizer, and fodder.[57] People also made extensive use of cryptomeria, felling it so liberally that first-growth specimens can be found today only in Akita, Yakushima, and a few other spots.[58] Then with a warmer and wetter climate beginning about 7500 BCE the land became more forested and grasslands declined, until a gentle cooling trend toward today's temperature levels set in about 2500 BCE. Based on pollen evidence, researchers believe agriculture came to Kyushu at the start of the Yayoi era (1000 BCE-300 CE), leading farmers to fell plains trees and eventually nearby forests. Much further east in Chiba prefecture, Japanese alders (*Alnus japonica*) were cleared for rice paddies by 600 CE.[59] With the rise of farming, peoples in Japan began to tame their natural environments and came to regard unwanted plant species as inimical, discovering the concept of weeds.[60]

Economic and political power in premodern Japan derived primarily from control over wet-rice agriculture and secondarily from dry-field farming, orchards, forestry, mining, and fisheries. Yet northeastern Japan had many cattle ranches during the Heian (794–1185) and medieval (1185–1600) periods, and the Kanto district had extensive horse pastures. These human interventions created a need for grasslands, which early farmers expanded by setting meadows on fire to engender new growth, provide high-quality fertilizer, and prevent reforestation. Meadows provided spaces for livestock to graze and farm hands to cut grasses for fodder or thatch.[61] Much as silviculture developed in the Edo period (1600–1868) to conserve Japan's forests, farmers in the same era subjected their grasslands to prudent use and careful conservation. In "Musashino" (1898), a story once read by every Japanese schoolchild, the novelist Kunikida Doppo (1871–1908) famously lamented that the Musashino plain west of Tokyo had lost its open meadows and was now an oak forest, the latest iteration of a region that was successively an ancient evergreen oak forest, a burned grassy prairie in the Heian period (794–1185), rice fields in the Edo era, and once again forested in the Meiji period (1868–1912).[62] Grassy vegetation abounded in open areas at Karuizawa, with its Nagakura horse pasture in the Heian period; on the southern slope of mount Fuji as well as the Bōsō and Izu peninsulas; and in Nagano prefecture, until commercial development encroached in the early twentieth century, with a resulting loss of the wildlife inhabiting these spaces. By one estimate, fields and meadows accounted for 13 percent of national land in 1900 but just 1 percent a century later,[63] because of industrialization, urbanization, population growth, and extensive plantation forestry. Newly replanted forests consisted of monotonous rows of cryptomeria, cypress, or larch to replace timber felled for urban construction and military materiel during World War Two.[64] Meanwhile chemical fertilizers, synthetic roofing, fossil-fuel heating, and a decline in livestock raising in the past half century have slashed the demand for grassland products. Without human management of *satoyama* and grassy meadows, one recent study asserts, "most of the plant life in the Japanese islands would turn into forests."[65] Although this warning may be alarmist, it is true that today forests cover about two thirds of Japan's national land—as in New England and other regions, a far greater proportion than in the eighteenth and nineteenth centuries.

In spite of global climate changes and the evolutionary imperatives of natural selection, it is astonishing that some species are considered virtually unchanged from 10 million years ago: ginkgos, the profusely flowering tall tree *Meliosma oldhamii,* Japanese chestnuts (*Castanea crenata*), willow-leafed magnolias (*Magnolia salicifolia*), and Japanese zelkovas (*Zelkova serrata*).[66] Visiting the Ōkunitama shrine in Tokyo's suburban Fuchū provides an opportunity to appreciate the 2,000-year history of this Shinto holy site, especially to learn that the long avenue of approach from the north is lined with ancient zelkovas donated by the Minamoto warrior family in 1062—and to realize that this species is essentially identical with its forebears in the Triassic period.[67]

How Japanese and Eastern North American Plants Came to Look Like Each Other

Today the floras of Japan and the North American Atlantic coast look remarkably like each other, but not just because plant collectors and commercial nurseries have imported species from Japan since the late nineteenth century. Deep geoclimatological forces such as glaciation, global warming, and continental drift have also driven the interchange of plant life from region to region across hundreds of millions of years, resulting in striking similarities of biota in both eastern North America and East Asia, where latitudes, topographies, land temperatures, and rainfall amounts are comparable. In addition to extinction events and ice ages, the formation and subsequent breakup of supercontinents were essential to the worldwide spread and later differentiation of plant genera that today remain surprisingly similar, in many cases, along the North American east coast and in Japan.

Paleogeographers and paleogeologists have employed the concept of supercontinents for more than a century to imagine the shape and distribution of earth's landmasses reaching back 3.6 billion years ago when Vaalbara, comprising regions in today's southern Africa and western Australia, may have formed a large continent lasting nearly a billion years. Martin Lewis and Kären Wigen have notably argued that metageographical concepts are constructed from cultural assumptions,[68] and this may hold true for speculation about the earliest supercontinents. But because geological evidence is fragmentary, it is unclear how such assumptions influence the geology of plate tectonics.[69] Of the seven major supercontinents hypothesized by scientists, all but the most recent had hazy configurations still subject to uncertainty and debate. Rodinia (1.2 billion years to 750 million years ago) is believed to have been a stark and hostile barren landmass lacking complex terrestrial life, surrounded by oceans filled with marine organisms.[70] Pannotia (620–560 million years ago), centered on Africa, seemingly embraced most of earth's current landforms, but only in 2018 did researchers establish a strong case for its existence, let alone its impact on animal or plant life.[71]

A so-called minor supercontinent, Gondwana (550–320 million years ago), was positioned mainly south of the equator and included about two thirds of earth's present-day land areas. South China and north China split off from Gondwana about 400 to 375 million years ago, the latter drifting toward the north pole, then later joining with Siberia, Mongolia, and south China within the next 100 million years.[72] Gondwana merged with Euramerica (also called Laurussia) about 335 to 330 million years ago to form the gigantic supercontinent Pangaea (335–173 million years ago), but it is still known to many scientists as Gondwana beyond the time Pangaea rifted into separate pieces more

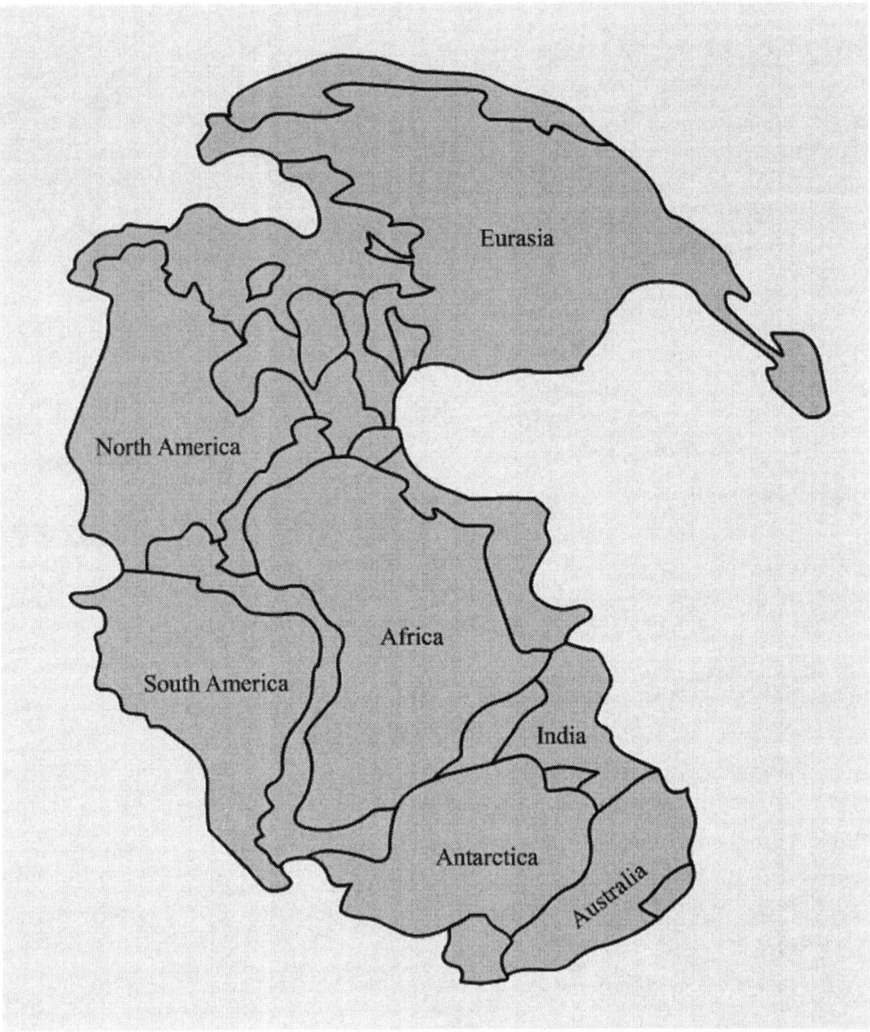

Figure 1.2 Map of Pangaea with modern continental outlines. En:User:Kieff.GNU Free Documentation License.

than 150 million years later. Pangaea witnessed gradual but substantial eco-diversity, including the development of vascular plants, gymnosperms, and early herbaceous plants. Its terrestrial plant life was bisected by the Great Extinction of 252 million years ago, which destroyed most terrestrial plant ecosystems yet had even more devastating effects on land animals. About 175 million years ago Pangaea finally broke in two, Laurasia in the north and Gondwana in the south, after which Gondwana gradually separated into

Africa, South America, India, Australia, and Antarctica. Like plants, the world's present continents are in constant motion. Even though today the Atlantic ocean widens by 2 or 3 centimeters per year, scientists speculate that 100 million years from now tectonic plate activity will shrink the ocean, and by 250 million years from now all seven current continents will merge into a new tropical supercontinent, dubbed Pangaea Proxima. In this projection Japan will rejoin the eastern edge of Asia, Tokyo will lie south of the equator, and Lagos will move more than 3,000 kilometers north of New York.[73] The effects of these movements on plant and animal life are virtually incalculable—and undoubtedly vast.

Laurasia (253–55 million years ago) consisted of today's North America and Eurasia, drifting north after the breakup of Pangaea and finally splitting into Eurasia (minus India) and North America (including Greenland) about 40 million years ago, but transit routes for exchanging plant and animal life between Europe and North America continued across a land bridge via Greenland until 25 to 15 million years ago.[74] (The biologist Robert Francis Scharff [1858–1934] once argued that a land bridge probably linked Europe with Greenland and North America as recently as 5 million years ago.[75]) The Beringian land bridge connected Siberia with Alaska at least as long ago as 58 million years but was most active in exchanging plants and animals starting about 23 million years ago, even during some ice ages, since the bridge was seldom glaciated because snowfalls were sparse. The Beringian land bridge was sundered about 5.5 million years ago, then revived intermittently between 135,000 and 30,000 years ago, at which point the land bridge emerged fully and persisted until rising seas caused by melting glaciers inundated it about 10,000 years ago.[76] In short, whether by land bridges, wind gusts, ocean currents, birds carrying seeds, or more recently humans in boats, plants had the means and opportunity to travel far from their points of origin, but to survive they needed favorable ecological conditions in new host environments. Eastern North America provided a welcoming locale for many floral genera from East Asia, and the reverse was sometimes true as well. What historian Christopher Parsons says of America and Europe is true a fortiori for eastern North America and East Asia: "Many of the species unique to American and European ecologies therefore share common ancestors and visible similarities but are genetically and morphologically distinct."[77]

In 1716 a French Jesuit, Joseph-François Lafitau (1681–1746), exploring near Montreal, discovered one widely distributed example of this hereditary affinity, American ginseng (*Panax quinquefolius*), which he recognized as closely related to Asian ginseng (*Panax ginseng*). Ginseng immediately became a booming business for merchants in New France. Amazingly, at its peak in 1752 an estimated 7.6 to 13.3 million ginseng plants were exported annually from North America to France, many of which were promptly reexported to China, where they fetched 180 francs per pound, versus 3 francs in New France.[78] Like Lafitau, the Swedish biologist Carl Linnaeus (1707–1778), considered the parent of modern taxonomy, was fascinated by the similarities of East Asian plants with those in eastern North America. His student Jonas P. Halenius (1727–1810) published *Plantae Rariores Camschatcenses* in 1750, considered the first scientific study to identify botanical affinities between East Asia (the Kamchatka peninsula) and the North American east coast. This volume, which enumerated nine species thought to exist in both places, was likely written by Linnaeus himself, as was usual at the time.[79] The Swedish botanist and physician Carl Peter Thunberg, a disciple of Linnaeus, served the Dutch outpost at Nagasaki for fifteen months between 1775 and 1776, long enough to collect plants around Nagasaki and along the way during a 750-kilometer journey to the shogun's capital in Edo (now Tokyo). These specimens formed the basis for his 1784 treatise *Flora Japonica,* which suggested that twenty plant groups found in Japan were also extant in eastern North America.[80] Many subsequent writers pointed out the botanical similarities between East Asia and eastern North America, including Luigi Castiglioni (1757–1832), an Italian linguist, travel writer, and commentator on the infant American republic.[81] The English botanist Thomas Nuttall (1786–1859), in *The Genera of North American Plants* (1818), likewise mentioned that many American species could also be found in Japan.[82]

Asa Gray (1810–1888), a medical doctor who apparently never practiced, taught natural history at Harvard University from 1842 to 1888 and became America's most important botanist of the nineteenth century. Conversant with Siebold's *Flora Japonica* (1835),[83] Gray produced a half dozen papers between 1840 and 1859 on the disjunct (separate) distribution of plants in East Asia and the North American east coast. In an 1856 preface to a catalogue of dried specimens collected by missionary-diplomat S. Wells Williams (1812–1884) and agronomist James H. Morrow (1820–1865) as a part of the 1853–1854 Perry expeditions to Japan, Gray noted similar or identical species in the two regions and pronounced one genus and forty species retrieved from Japan as new to science.[84] He enthused, "No part of the world beyond his own country offers, as to its vegetation, a greater interest to the botanist of the United States than Japan."[85] Gray's observation remains surprisingly applicable today, especially among horticulturalists and landscape architects around the globe.

Gray likewise studied the plants collected in 1855 at Shimoda and Hakodate by Charles Wright (1811–1885) during a North Pacific surveying expedition,[86] concluding in "a revolutionary paper"[87] published in 1859 that about 580 Japanese species were "allies" with those in the eastern United States or other north temperate zones. He found that along the North American east coast, 61 percent of the Japanese taxa were present, and 23 percent of species were identical, very likely of common descent rather than separate in origin.[88] The affinities he found in flowering plants existed also for fungi, insects, and other life forms. Gray's analysis of Wright's specimens provided detailed evidence of connections between the two floristic zones that had long been hypothesized but never before proved. In an obituary of Gray, Charles Sprague Sargent (1841–1927), first director of Boston's Arnold Arboretum, called Gray's 1859 paper his "most remarkable and interesting contribution to science."[89] Charles Darwin's letters to Gray in 1856 indicate that Darwin was aware of the same plant genera in eastern North America and East Asia; Gray's concrete findings mainly supported Darwin's evolutionary theory and refuted the polygenistic views of Harvard naturalist Louis Agassiz (1807–1873), who rejected evolution and believed that human races were created separately.[90]

Gray's scholarship on East Asian and North American plants inspired further research during the next century and more, culminating in international conferences held in 1969, 1982, and 1983. Sargent built on Gray's work to compile statistics on Japanese flora, noting that in many respects vegetation in Japan was quite dissimilar to that in eastern North America. After ten weeks in Japan, Sargent rather impressionistically declared that the deciduous trees of the North American east coast were "more beautiful" than their Japanese counterparts.[91] Between 1867 and 1871 Dutch botanist Friedrich A. W. Miquel (1811–1871) rebutted Gray's ideas of common descent, arguing that 81 species shared between East Asia and the United States emerged independently.[92] More recent studies have dismissed Miquel's views of separate origination; a scholar of vascular flowers from East Asia and the Blue Ridge mountains of the southeastern United States reported that "one might even argue for cogeneric status" for his own specimens.[93] Gray's theory that glaciers advancing from the Arctic caused flora to move south has been superseded by research showing that different plant groups have diverse histories stemming from factors such as climate, geography, natural selection, and speciation,[94] yet his analysis showing kinship between East Asian and eastern North American plant groups continues to inform investigations by specialists on both sides of the Pacific.

Gray and many other botanists believed that North American plants with relatives in East Asia arrived via land bridges across the Bering sea. Michael Donoghue and Stephen Smith, writing in 2004, found that most of the temperate forest plant groups with counterparts in East Asia in their research sample moved out of East Asia to North America during the last 30 million years, often after the last North Atlantic land bridge disappeared 25 to 15 million years ago: "This favours Beringia over a North Atlantic land bridge as the primary path."[95] Richard Ian Milne concurred, reasoning that many plants moved to eastern North America after North Atlantic land bridges broke apart, so had to enter via Beringia, although East Asia was not necessarily their point of origin.[96]

Yet recent studies have also highlighted two routes across the North Atlantic from Europe as probable modes of transmission, one bridging Greenland and Scandinavia, the other Greenland and southwestern Europe. For early epochs there is considerable evidence of genetic exchange across the Atlantic. David Boufford, a specialist on systematics, concluded that the Bering land connection was not the key route of transmission that Gray thought; paleobotany and plate tectonics show that many Asian plants came to North America from Europe when the two were close enough for easy interaction.[97] Biologist Steven Manchester pointed to fruit and seed taxa identified from London, the Rocky mountains, and the Great Plains as proof of a "major dispersal" between North America and Western Europe between 66 and 50 million years ago. The next 10 million years saw a sizable influx of Asian plants, many of them via Europe, then increasingly arriving across the Bering land bridge after about 40 million years ago.[98] Although fossils tell little about the direction of plant movements, they confirm that certain genera now extinct in western North America because of drought survive today in eastern North America or East Asia.[99] They also show that a number of plants found in Japan spread as far as northern Latin America, including such genera as magnolias and deutzias, both of which are also found in the Himalaya.[100]

While acknowledging the limits of both fossils and molecular data for estimating when plant groups moved to new biotas, evolutionary biologist Jun Wen posited that various angiosperms from East Asia may have crossed the Atlantic to eastern North America sometime between 66 and 34 million years ago, a time when many plant species lived throughout the globe in the north temperate zone. Examining forests in East Asia and North America, Wen found that deciduous species probably advanced across both the Bering and the North Atlantic land bridges until the latter were closed off by continental separation; an Aleutian land bridge may also have existed, allowing flora to cross in both directions.[101] Paleobotanist Bruce Tiffney concluded that tropical evergreens and deciduous trees probably crossed to North America from Europe during the global warmth of 56–50 million years ago, after which the Bering land bridge became increasingly active in cross exchanges of temperate species as the earth gradually cooled. Tiffney found no evidence that East Asia was the sole origin of today's plant life in the northern hemisphere; the floral similarity between East Asia and eastern North America developed not from "a single historical event but is the result of a layering of many events"[102] across millions of years, seemingly on a species-by-species basis starting with maples and pachysandra.[103] As the influx of East Asian plant life to North America waned after about 5 million years ago, and as repeated ice cycles imperiled certain species in North America, more and more disjunct plant communities emerged, evolving separate species from the same genera found in the two regions. Yet Tiffney cautioned, "We may safely assume that the eastern Asian-eastern North American pattern among evergreen or thermophilic taxa arose at a time when these plants could move directly between the two areas and is not a result of parallel evolution from widespread, deciduous temperate common ancestors."[104]

Most surviving plant genera in North America trace to the Tertiary period (66–2.6 million years ago), many of them angiosperms shared with East Asia and Europe during the global warmth between 66 and 34 million years ago. Some subsequently disappeared from Europe and North America and are now found only in East Asia. Conifers and other alpine trees were increasingly common during the cooler Miocene epoch (23–5.3 million years ago) in all three regions.[105] During the past three decades collaborative fieldwork between East Asian and American botanists, using molecular methods and cytological information, has refined earlier morphological studies of similar plants in East Asia and eastern North America.[106] Nearly a century ago a detailed study of plants and trees in Hokkaido identified 435 species also found in North America, representing 26.7 percent of all species in Hokkaido.[107] Phytogeographer Hara Hiroshi (1911–1986) collected skunk cabbage (*Simplocarpus foetidus*) during 1939–1940 in Massachusetts and in Shimotsuke, Tochigi prefecture, finding that the two samples agreed "perfectly" in most characteristics and were the same species, although the Japanese version was "a variety of the American" because it had broader leaves and a less fetid odor.[108] At the same time Hara compiled a list of 456 plant species he identified as common to Japan and North America, of which 354 were present along the eastern seaboard and the rest elsewhere in North America.[109]

Two decades later bryophytologist (specialist on mosses) Iwatsuki Zennosuke (1929–2015) reported that about 10 percent of southern Appalachian mosses were found only in eastern North America and in Japan or were closely related to Asian species, all growing in moist ledges or shady ravines. He detailed the similarities of two species of the genus *Fissidens*, concluding that they were essentially the same in Japan as in the Appalachians.[110] In 1974 botanist Hotta Mitsuru (1935–2015) offered a list of 13 genera with species "native" to both Japan and the American east coast, including 30 species of magnolias in Japan and 8 in the U.S., 6 Japanese and 3 American species of witch hazel, and a single species of lotus in each site, based, like Hara's and Iwatsuki's work, on morphological similarities.[111] Iwatsuki and two colleagues refined previous research in a 1994 study emphasizing commonality of species with Japan mainly among plants living in American coniferous forest floors, alpine locations, and salt marshes.[112]

How many plant genera exist in East Asia and eastern North America is still debated among specialists, with estimates ranging from 65 to at least 120. Geobotanist Wu Zhengyi (1916–2013) identified more than 120, of which 117 lived in China, comprising 4 percent of the 2,980 higher plant genera in that country, mainly temperate plants descended from ancestors that failed to survive ice cycles in Europe.[113] A recent compilation found more than 90 genera of plants, including pairs of counterpart species (*vicariads*), native to both Japan and the southeastern United States,[114] although phylogenetic analyses have cast doubt on how close so-called sister pairs are to each other.[115] Biologist Robert Askins notes that deciduous forests in Europe, eastern North America, and East Asia had common origins but later evolved in isolation for millions of years. Fossils show that genera now extinct in Europe are alive in North America; others, such as ginkgos, disappeared from both Europe and North America but persisted in East Asia and developed new species. Askins counts about 65 plant genera currently found in eastern North America and East Asia, including such familiar vegetation as witch hazel, catalpa, and Virginia creeper.[116] Jun Wen also cites 65 genera shared by eastern North America and East Asia, left over (relict) from vegetation distributed globally in north temperate regions between 66 and 23 million years ago, with separation into distinct ecologies in the two regions between 23 and 5.3 million years ago. Most surviving plant groups are woody, many showing almost no morphological change after separation from counterpart species across the Pacific.[117] In short, although the number of genera shared by East Asia and eastern North America may be in doubt, there is no question that the commonalities first identified by Lafitau, Linnaeus, and Thunberg, and the similarities first proved by Asa Gray, have won near-universal acceptance by scientists and horticulturalists around the world.

Less celebrated, but no less intriguing, are affinities between plant groups in East Asia and western North America. A North American interior seaway formed between 100 and 66 million years ago, creating a tropical body of water stretching from the Gulf of Mexico to the Arctic Ocean that split the continent in two. The seaway retreated during the Paleocene epoch (66–56 million years ago) because the Rocky mountains uplifted, resulting in a dry climate to the west that caused many broadleaf forests to become extinct. In their place arose deserts, grasslands, or conifers in higher elevations.[118] More than a century ago plant collector Ernest Henry Wilson (1876–1930) enumerated 13 genera of pines in North America and 12 in Japan. Of the 32 species and 8 varieties in Japan, he identified only 12 species and 2 varieties that existed naturally outside Japan. Wilson concluded that conifers in Japan were much more akin to those in western than eastern North America, except for hemlocks.[119] In the 1970s Hotta Mitsuru found that Asian skunk cabbage (*Lysichiton camtschatcensis*) thrived both on the snowy Japan sea coast and along milder parts of the North American west coast but not in eastern North America.[120] Some ferns that live in Japan exist in both eastern and western North America, as is true of the *Liquidambar* (sweetgum) genus and the *Aesculus* genus, which includes chestnuts and buckeyes. Fossils as well as phylogenetic and biogeographic studies of East Asia and both regions of North America have brought out complex relations among plant groups well beyond what Asa Gray and other pioneers posited more than 150 years ago.[121] Nonetheless, it is irrepressibly obvious that plants have been in motion for millions of years, undeterred by the breakup of supercontinents or the collapse of land bridges—and greatly abetted by human interventions in modern times.

Two

Plants in Early Modern Japan

Japanese began the Edo period (1600–1868) more often in awe or fear of the natural world than in fond sympathy, let alone romantic harmony, with their nonhuman surroundings. Even as they exploited timber for dwellings and grain stalks for thatch, most Japanese understood droughts, floods, plagues, typhoons, earthquakes, and other natural phenomena through some combination of religious and superstitious beliefs. Yet they ended this era with a firm awareness of plants, animals, and their own place in the natural environment they shared with living organisms and inanimate objects. This chapter details the process whereby Japanese arrived at a sophisticated knowledge of the plant life surrounding them, through successive understandings fully commensurate with, but distinct from, their counterparts in Europe. It shows that early modern science, although much affected by Chinese precedents, developed along lines specific to Japan's own ecology—but without achieving a scientific revolution in either method or result before the full-scale arrival of Western physical and life sciences in the later nineteenth century.

A century ago the Japanese botanist Shirai Mitsutarō (1863–1932) divided the history of plant science in his country into three overlapping phases: (1) the study of medicinal plants, chiefly from Chinese and Korean books on materia medica (herbal medicine, Ch. *bencao,* Jp. *honzō*), throughout the Edo period; (2) a focus on natural history in the later eighteenth century, mainly by academy-based professionals who identified, described, and illustrated a full range of Japanese plants from wild or cultivated specimens; and (3) an early modern penchant for systematizing the accumulated knowledge of plants by classifying them according to Linnaean or later European methods, thus laying the foundations of modern biology after 1868.[1] With modifications, these phases continue to guide writings today on how Japanese understood plants in the Edo era, while also recognizing the impact of such "Dutch doctors" as Engelbert Kaempfer, Carl Peter Thunberg, and Philipp Franz von Siebold in spreading European botanical knowledge in Japan from their outposts as physicians serving Dutch traders at Nagasaki during the Edo period. Collectively these phases of investigation and inputs from abroad transformed how Japanese perceived plants in the early modern era.

Chinese Herbal Medicine in Japan

Today specialists in Japan count more than one thousand medicinal plants, 148 of which are medicines approved for reimbursement under National Health Insurance. Another 1,370 herbal substances are banned as of 2013 under the Pharmaceutical Affairs Law as poisons or impairments to driving.[2] Nearly all herbs legal today—some stemming from India's ancient Ayurveda medications—were introduced centuries ago by Chinese and Korean practitioners, including ginkgo seeds used for longevity, fertility, and memory.[3] Chinese knowledge of herbal medicine traces to the *Erya* (3rd century BCE), a lexicon of the classics that included many plant names, and botanist Ji Han's (263–307 CE) *Nanfang caomu zhuang* (c. 304), a compendium of subtropical plants in southern China and northern Vietnam. Chinese medicine came to Japan's Yamato court as early as 415 CE when a doctor from Korea treated the Ingyō emperor (r. 412–453). The Tang dynasty (618–907) pharmacopeia *Xinxiu bencao* (20 vols., 659), imported to Japan by a student returning from China, was used in Japanese medical education at the State Medical Office (Kusuri no Tsukasa) and Imperial University (Daigakuryō), both dating to 701, along with tea, ginseng, herbs, roots, spices and other concoctions brought by Chinese merchants.[4] At the same time the distinctly nativist *Kojiki* (Record of ancient matters, 712), Japan's oldest extant book, included references to indigenous herbs used in folk medicine, as did *Izumo fudoki* (Izumo gazetteer, 733).[5] Japan's earliest literary works, such as *Man'yōshū* (Collection of ten thousand leaves, 759), *Kokinshū* (Ancient and modern collection, 905), and *Genji monogatari* (Tale of Genji, 1008), contain profuse references to flowers that have fascinated researchers for many centuries. Scholarly databases, academic libraries, and major bookstores abound with titles analyzing the botany as well as the symbolism of the plant life surrounding characters in these early creative works.[6] In 2015 Japan had 37 Man'yō botanical gardens, all but ten of them open to the public, although most visitors probably came to enjoy the vegetation rather than the literary allusions, despite ample signage with poems cued to the flowers on display.[7]

Honzō wamyō (Japanese names for medicinal herbs), compiled in 918 in eighteen volumes by Fukane Sukehito (Fukae Hōjin, 898–922), was Japan's first book of Chinese drugs, enumerating 1,025 species of materia medica with Japanese names for each. Fukane's work was soon followed by *Wamyō ruijushō* (Chinese-Japanese dictionary, 934), a twenty-volume encyclopedia prepared by Minamoto Shitagō (911–983) summarizing biological knowledge up to that time. Another key text was *Honzō irohashō* (Medicinal herbs in order, 1284) by physician Koremune Tomotoshi (fl. late thirteenth century), who commented on the difficulties Japan faced in producing medicines mentioned in Chinese books.[8] These East Asian works and the herbal drugs they cataloged predated the rise of materia medica in Europe, where the fifteenth-century printer Johann Veldener (fl. 1473–1486) compiled *Den Herbarius in Dietsche* at Leuven, Belgium, in 1484. The first professor of herbal medicine was appointed at the Papal University of Rome in 1514, supported by a garden for growing medicinal plants as a laboratory for students' work. Soon other European universities followed suit, and eventually the arrival of Jesuit missionaries in 1549 brought European knowledge of materia medica to Japan.[9]

But far more significant was *Bencao gangmu* (Outline of medical herbs, Jp. *Honzō kōmoku*), compiled by Chinese naturalist and physician Li Shizhen (1518–1593) and published posthumously in 1596. This encyclopedic work, which drew on Confucian histories and Daoist texts dating to the first century CE, described how to harvest animals and plants for medicinal use and listed 16 groups comprising 1,892 drugs considered efficacious. Arriving in Japan in 1607 as the new Tokugawa state was embracing Neo-Confucian learning, *Bencao gangmu* became the basis for herbal medicine for the rest of the Edo period, establishing morphological resemblances through the study of plant forms and structures as the key to classifying vegetation.[10] Hayashi Razan (1583–1657), a Neo-Confucian philosopher and tutor to four Tokugawa shoguns, bought a copy in Nagasaki and presented it to the retired shogun Ieyasu (1542–1616). In his twilight years Ieyasu was perhaps the first important patron of secular (non-Buddhist) professional researchers, whose task was to grapple with the flood of Chinese texts imported in the seventeenth century.[11]

Europeans Embrace Japanese Plants

While Japanese physicians and other researchers immersed themselves in *Bencao gangmu* and similar works from China, far on the periphery of Japan a few specialists tried to make sense of European science through contact with visitors such as Luís de Almeida (1525–1583), a Portuguese medical missionary who arrived in Japan in about 1555 and spent the rest of his life in Kyushu, teaching Western medicine and building Japan's first hospital. A century later the Dutch East India Company sent Willem ten Rhijne (1649–1700) to Japan from 1674 to 1676 at the request of shogun Ietsuna (1641–1680, r. 1651–1680), who sought a physician with botanical and chemical knowledge.[12] The European botany that ten Rhijne and others introduced traced its roots to the Greek polymath Theophrastus (371–287 BCE), whose nine-volume *Enquiry into Plants* and six-volume *On the Causes of Plants* are considered germinal. Equally influential was Pliny the Elder (23–79 CE), a Roman officer and naturalist who ironically died when mount Vesuvius erupted, just two years after he completed *Naturalis Historia* in thirty-seven books.[13] Early modern botany dates to the late fifteenth century in Europe, much of it concerned with economically useful plants but also with the exotica that princes and rich merchants prized for ornamenting their gardens. Johann W. von Kaub (1430–1504) completed *Gart der Gesundheit* (Garden of health) at Mainz in 1485, considered Europe's first herbal manual, a volume describing 382 types with drawings of plants from Germany, Italy, and Jerusalem.[14] Like the Chinese herbal medicine then popular in Japan, European study of plants was considered a part of medicine until the mid-seventeenth century, at which point the term *botany* emerged to describe the newly independent discipline of examining "useless" plants.[15]

Two Germans employed by the Dutch East India Company were the first Europeans known to express deep interest in Japanese plants. Andreas Cleyer (1634–1698) was a physician who twice served as *kapitan* (a Japanese term for chief trade official) at the Dutch enclave in Nagasaki between 1682 and 1686, then was evicted by the Japanese for supposedly failing to control smuggling.[16] Beginning in 1686 Cleyer wrote at least 29 letters (*observatii*) to scholars in Europe explaining the vegetation he found in Japan, including descriptions and illustrations of 53 plants sent to the elector of Brandenburg. After returning with Cleyer to Dresden in 1689, his gardener and assistant, George Meister (1653–1713), described eighty-nine Japanese plants—including dwarf potted plants and *Camellia japonica*—in *Der Orientalisch-Indianische Kunst- und Lust-Gärtner* (Oriental-Indian art and pleasure gardener, 1692). Cleyer's and Meister's writings are considered the earliest European reports on Japanese plant life, although they focused mainly on medical herbs and had limited impact.[17]

Far more influential, if scarcely more adept scientifically, was the German physician Engelbert Kaempfer (1651–1716), who ministered to the Dutch settlement at Nagasaki between 1690 and 1692. Kaempfer was insatiably curious about Japan and a tireless cheerleader for the beauty of its plants: "Japan I think may vie with most, if not all, known Countries, for a great variety of beautiful plant and flowers, wherewith kind nature hath most liberally and curiously adorn'd its hills, woods and forests."[18] Kaempfer built a small botanical garden at Deshima, the artificial island in Nagasaki's Urakami river where Dutch traders were confined during the Edo period. He experimented with numerous species and provided Western-language names and clear illustrations for many of them, including ginkgo, a tree until then unknown in Europe. His aesthetic fascination with plants must have seemed normal to his Japanese associates, because most medications in that country were herbal; he boasted that he plied informants with a "cordial and plentiful supply of European liquors" in exchange for details about local plants.[19]

In 1712 Kaempfer published *Amoenitates Exoticae* (Exotic pleasures), a sprawling account of his travels throughout Persia, Southeast Asia, and Japan. The fifth fascicle (volume), titled *Flora Japonica,* drew on *Kinmō zui,* a twenty-volume illustrated encyclopedia compiled in 1666 by Confucian scholar Nakamura Tekisai (1629–1702), the final five volumes of which dealt at length with Japanese plants.[20] Kaempfer's *Flora* included 256 plant sketches and listed 324 species recognized today, a number of which originated in China. As with Meister's book twenty years earlier, *Camellia japonica* figured prominently, as did two species of the *Hosta* genus.[21] Carl Linnaeus admired Kaempfer and relied extensively on his descriptions of plants encountered during his travels. Other botanists honored Kaempfer by naming various species *Kaempferi,* including Japanese larch (*Larix kaempferi*).[22] Kaempfer's *History of Japan* (1727), extensively edited and translated into English by Swiss librarian Johann Caspar Scheuchzer, quickly became a best seller in a Europe suddenly eager to learn about East Asia, appearing in twelve editions and translations within a decade. Much more than a mere catalog of vegetation, Kaempfer's *History* commented extensively on useful plants such as mulberry for silk, bamboo for construction, and the bark of the paper mulberry (*Broussonetia papyrifera*) for clothes as well as paper. It described Japan's main agricultural crops and many varieties of mushrooms, noting that "people often die of them."[23] Volume three commented at length on the history of tea plants in Japan, how the leaves were harvested and roasted, and the proper way to drink tea.[24] The book may have been far more successful in a scholarly sense than previous Western writings on Japan, with an undeniable impact on European knowledge of that country for the next century,[25] but part of its appeal was surely its breezy tone and pithy observations about Japanese customs and practices previously unknown in Europe.

Natural History and Indigenous Plants

Chinese herbal medicine remained the chief focus of Japanese scholars interested in plants throughout the eighteenth century, but a parallel track called natural history (*hakubutsugaku*) gradually developed early in the century and grew robust after 1750. Natural history, in Japan and elsewhere, was concerned with the entire world of nature—and with all forms of vegetation, no longer merely medicinal herbs. Natural history in the West stemmed from ancient Greece and Rome, both of which produced works on flora and fauna. Florentine polymath Leonardo da Vinci (1452–1519) was fascinated by geology and believed science and technology were intimately tied to the visual arts. Philosopher and statesman Francis Bacon (1561–1626), sometimes called the parent of inductive scientific inquiry, took the nonhuman world as an object of research separate from humanity in his *Historia Naturalis,* posthumously published in 1638. This work called for impartial investigation via rigorous experiments without reference to religion, even though Bacon was a staunch Anglican. As objects from afar flooded Europe via exploration and commerce after 1700, naturalists collected, described, and classified specimens from the global animate and inanimate realms,[26] "attuned to the notion that the truth of the natural world could be discerned through disciplined observation."[27] Natural history was less anthropocentric than herbal medicine, which focused specifically on human well-being; by the eighteenth century natural historians institutionalized their studies in botanical gardens and herbaria.[28] These specific developments in Europe had little immediate impact in Japan, but more and more scholars trained in *honzō* expanded their compass to the country's entire flora as natural history became an established area of inquiry in the eighteenth century, separate from but in chronological juxtaposition with counterparts in the West.

Herbal medicine in East Asia was practical in objective and instrumental in application: relieving discomfort, remedying illness, enhancing nutrition through plant crops. Japan's natural historians, in search of pure knowledge, began writing about every variety of vegetation surrounding them, in tandem with the shift from Kyoto to Edo as the main site of research by the mid-eighteenth century. Ming and Qing works of natural history now accompanied those on materia medica in Japanese researchers' libraries. Two works written in the early eighteenth century illustrate the gradual shift to investigating Japan's indigenous flora as well as refining understandings of Chinese herbal medicine. Kaibara Ekiken (1630–1714), a Neo-Confucian philosopher and preeminent intellectual of his age, completed *Yamato honzō* (Japanese herbal medicines) in 1709—"the first genuinely Japanese encyclopedia of natural history," according to historian Federico Marcon.[29] Kaibara Ekiken, an inveterate traveler and plant collector, is sometimes regarded as the founder of native herbal studies because he tried to Japanize the China-centered materia medica then regnant in Japan. *Yamato honzō* glossed 1,362 species in 37 genera, all of them known in Japan in his day, including 772 listed in *Bencao gangmu,* 203 cited in other works, 29 imported from the West, and 358 species of medicinal plants and animals considered unique to Japan.[30] Six years later the painter and naturalist Inō Jakusui (1655–1715) finished 362 of a planned 1,000 volumes of *Shobutsu ruisan* (Classification of things) before his death, a project completed by Niwa Shōhaku (1691–1756) by 1745.[31] *Shobutsu ruisan* sought to synthesize Chinese texts on herbal medicine and other natural subjects as a guide to nomenclature for plants and animals in Japan; based on his regional surveys for the Tokugawa regime, Niwa added many new species not found in China.[32] Both works imposed strict methods of classifying knowledge, and both emphasized empirical applications of learning for the benefit of other scholars wrestling with the massive bulk of Chinese scientific works now available in Japan. By the end of the Edo period, according to one estimate, Japanese specialists had produced 150 major works of natural history and another 600 minor ones, all with some botanical content.[33] In the twentieth century natural history (*hakubutsugaku*) lost favor with scholars of modern botany, although most plant ecologists continued to combine natural history with rigorous survey and experimental methods. The spirit of natural history persisted among amateur naturalists through botanical fan clubs (*dōkōkai*) and magazines. Today general investigations of natural phenomena are usually called *shizengaku* (studies of nature) and are often conducted by knowledgeable laypeople.[34]

One of the greatest patrons of Edo-period scholarship was also one of the most remarkable shoguns in Japanese history, Tokugawa Yoshimune (1684–1751, r. 1716–1745). He hired scholars of herbal medicine as government employees, distinct from the amateur naturalists who proliferated later in the century.[35] Pragmatic and in some ways progressive, Yoshimune adopted the import substitution policies of Arai Hakuseki (1657–1725), a conservative adviser who sought to grow domestic medicinal herbs and other products to offset a crisis over exporting gold and silver from Japan. Matsuoka Joan (1688–1746), a pupil of Inō Jakusui, came from Kyoto to lecture the shogun about materia medica; in 1722 Niwa Shōhaku helped form a committee of twenty-five wholesalers to vet Japan-grown medicinal herbs in Edo, establishing a degree of medical independence from China. In 1717 Yoshimune decreed that the government's medicine garden be expanded tenfold to grow ginseng domestically, and in 1719 the government began transplanting ginseng smuggled from Korea, a project that met with no success until 1728, but by the end of Yoshimune's rule ginseng roots were found throughout the country, and like North American growers the Japanese were exporting roots to China by the 1780s.[36] More recently the tables are turned: Japan has often had to import herbal medicines from China because most of the three hundred frequently used herbs are too difficult to grow domestically.[37]

The shogunal government's herb garden began in 1638 in two locations north and south of Edo castle and was eventually consolidated at Koishikawa in 1711; soon thereafter each feudal domain had its own garden. Yoshimune ordered the gardeners at Koishikawa to provide medicines to the poor starting in 1722, and it soon became a de facto neighborhood pharmacy. Three years later the Dutch contributed oak shoots and eighteen medicinal plants from Europe, as well as seeds, for cultivation at Koishikawa.[38] The first sweet potatoes in the Kanto region near Edo were apparently grown there in 1735. After the Meiji restoration of 1868 the shogunal garden was briefly administered by the city of Tokyo as successor to Edo, then became the garden of the newly founded Medical College (Igakkō) the following year. When the University of Tokyo was established in 1877, it assumed control of the renamed Koishikawa Botanical Gardens. Today the facility covers 16 hectares and contains 5,000 plant specimens from Japan and East Asia, including specimens of 2,000 tropical and subtropical species in its greenhouses, but it no longer serves as a primary source of healing medicines.[39]

Yoshimune ordered overlords (daimyo) to catalog the natural products in their domains and send the results to his headquarters in Edo, a clear instance of making society legible, in James Scott's terms, by rearranging governmental land controls.[40] He commissioned Niwa Shōhaku to supervise nationwide surveys between 1734 and 1736 in order to step up agricultural output through more effective exploitation of resources in the wake of the Kyōhō famine of 1729–1735, a step that led to greater commercialization of farm output.[41] Three generations later the political economist and budding Pan-Asianist Satō Nobuhiro (1769–1850) advocated agricultural development to bolster the realm in the face of potential outside threats. Botanists in a number of domains engaged in self-strengthening projects to exploit natural resources to human advantage. The southwestern domain of Satsuma (part of today's Kagoshima prefecture) was notable for its economic reforms that in some sense paved the way for the mercantilism of the post-1868 Japanese state.[42]

Yoshimune also sponsored a Japanese-language translation of *Cruijdeboeck* (Book of plants), published at Antwerp in 1554 by the Flemish physician and botanist Rembert Dodoens (1517–1585), with 715 images, mainly of medicinal plants. An expanded edition of this massive volume was introduced to Japan about 1650 but gained little attention until Yoshimune opened a copy at Edo castle in 1717. Eventually he directed Noro Genjō (1693–1761), a specialist on materia medica who had studied with Inō Jakusui, to head a team to translate Dodoens' book into Japanese. The project spanned the years from 1742 to 1750 and produced *Oranda honzō wage* (Dutch herbal medicine translated) in eight volumes that constituted a highly abridged adaptation, discussing just 106 Western plants, mainly medicinal herbs.[43] Chinese and Japanese equivalents were supplied for Dutch and Latin plant names. A prominent daimyo who served as chief councilor to the Tokugawa government between 1787 and 1793, Matsudaira Sadanobu (1759–1829), harbored an aesthetic fascination with Western natural history and ordered Ishii Shōsuke (1743–?) to prepare a complete Japanese-language translation of *Cruijdeboeck*. The project occupied thirty years and ran to 170 manuscript volumes, but it was never brought to print, despite Matsudaira's efforts beginning in 1823 to find a publisher. Many volumes and engravings were lost in the Edo fire of 1829, and only portions of the manuscript survive, including a copy in the Waseda University library. Early in the nineteenth century Yoshio Gonnosuke (1785–1831) prepared an extensively illustrated adaptation of the 1644 edition of *Cruijdeboeck* but included only 84 plants.[44] These and other works show the continuing grip of Dodoens' volume, even as Japanese plant specialists transcended herbal studies and plunged into Western scientific works (*yōgaku*) later in the Edo era.

After the midpoint of the Edo era (1730), Japanese perceptions of plants gradually shifted from philological and pharmacological to descriptive and visible, in both word and image, embracing the universe of Japanese flora and to some degree those abroad. A number of scholars trained in materia medica took a deep interest in natural history, sometimes conducting fieldwork in farms and forests, while amateurs collected and cultivated wild plants for their novelty and eye appeal.[45] At Edo castle Tamura Ransui (Gen'yū, 1718–1776), a specialist in herbal medicine, advised Yoshimune on ginseng and wrote *Ninjin kōsakuki* (Cultivating ginseng) in 1747. Based in the shogunal stronghold, he was little affected by studies of medicinal herbs in Kyoto and bent his energies toward raising agricultural output. Together with his star pupil, the physician and painter Hiraga Gennai (1728–1780), Tamura organized Japan's first nationwide nature fair in Edo in 1757, making hundreds of rare plants and animals visible as never before. Four more exhibits followed between 1758 and 1762, the latter displaying 1,300 specimens from Japan, Southeast Asia, and further afield, mainly assembled from Japanese herbalists and other private collectors.[46] Hiraga Gennai had studied both herbal medicine and Western learning, and once the nature fairs came to an end he produced *Butsuri hinshitsu* (Qualities of natural products, 1763), which examined 360 natural materials classified according to Li Shizhen's *Bencao gangmu.* Included were about 40 "barbarian" (Western) products, reflecting knowledge available to him through the Dutch language; excluded were curiosities of no practical impact. The comparative literature scholar Haga Tōru has called Hiraga's book "an attempt to unsettle traditional authorities of the China-oriented school of herbals" in Japan.[47] In addition to his success as a naturalist, Hiraga gained fame as a playwright, satirist, and mine operator—a polymath in an age of increasing specialization.

In Kyoto the herbal pharmacologist Matsuoka Joan (Gentatsu, 1669–1747) served as the pivot from the earlier scholarship of his teacher Inō Jakusui to Matsuoka's influential pupil Ono Ranzan (1729–1810), who was an almost exact contemporary of Hiraga Gennai. Matsuoka wrote extensively on medicinal and other plants but is recognized today mainly through the achievements of his students.[48] Of these, Ono Ranzan was one of the most celebrated, for his mastery of both Chinese herbal remedies and Western medicine. He taught herbal pharmacology to a thousand or more pupils at a school he opened at the age of twenty-five in Kyoto; much later he joined the shogunal Institute of Medicine (Igakukan, est. 1791) in Edo and conducted surveys of herbal drugs throughout the country from 1799 until he died in 1810. Ono's field-work suffused his forty-eight-volume *Honzō kōmoku keimō* (Commentaries on *Bencao gangmu,* 1803), which was perhaps the most highly regarded of the many hundreds of works on herbal medicine of his era. His analyses of Japan's natural history in this book were considered detailed, accurate, and bold in advocating experimentation rather than repeating truisms from Chinese medicine.[49] Like Hiraga, Ono paid close attention to native Japanese plants and brought a high degree of empiricism, partly informed by his knowledge of Western medicine, to his research.

Scientific Illustrations

Botanic illustrations were artistically elegant and scientifically accurate representations that made plants and their flowers visible to the political elites for whom they were carved on woodblocks and painted by hand in watercolors. Federico Marcon has pointed out that the proliferation of illustrated manuals in the later Edo period prompted a shift toward inductive observation of reality, shelving earlier Neo-Confucian assumptions that all knowledge was a priori, merely to be rediscovered by each successive generation. This shift was domestically generated within Japan, not the result of Western science,[50] although the full embrace of inductive logic by Japanese scholars awaited the later nineteenth century.[51] Natural history drawings in the Edo period were noncommercial, produced in limited editions and often intended to be works of art as well as faithful reproductions of biological realities found in individual species. Unlike colonial visuality in the same era in Spain, which made far-flung colonies visible at home through illustrations of plants,[52] Japan's was an internal visuality, making known the flora and fauna of the 260 domains of the shogun's realm.

An early example of botanical drawings was the work of the prolific illustrator and painter Tachibana Morikuni (Nakamura Yuzeishi, 1679–1748), whose *Kusubana ōshukubai* (Picture book of flowering plants and the nightingale in the plum tree, 1727f) has been reprinted a number of times.[53] Exquisite images are found in *Honzō zufu* (Illustrated materia medica) by shogunal botanist and zoologist Iwasaki Kan'en (Tsunemasa, 1786–1842), whose hand-painted woodblock prints eventually occupied ninety-three volumes issued between 1818 and 1921. Iwasaki's scholarly credentials were impeccable: he was a pupil of Ono Ranzan, published an illustrated natural history of the Kanto region near Edo, and later became associate director of the Koishikawa

Botanical Gardens.[54] Mizutani Hōbun (Toyobumi, 1779–1833), a student of Western medicine who coedited a manual of flora and fauna titled *Buppin shikimei* (Names of things clarified, 1809), later compiled *Honzō shashin* (Images of medicinal herbs, 1826), featuring drawings of seventeen specimens, annotated with Linnaean plant names. Mizutani also founded the herbalist association Shōhyakusha in 1828 and served as director of the Higashiyama Botanical Garden in Nagoya.[55]

Another illustrator was Kawahara Keiga (1786–1862), the only Japanese artist permitted to come and go at the Dutch settlement in Nagasaki between 1811 and 1842. He sketched the illustrations for Philipp Franz von Siebold's *Flora Japonica* (1835, 1870), although in many cases European artists produced

Figure 2.1 Kawahara Keiga (1786–1862), *Pomegranate* (*Punica granatum*). National Museum of Ethnology, Leiden. Public domain.

the final versions from Kawahara's models.[56] A talented painter whose work is displayed today in Japanese museums, Kawahara was imprisoned in 1829 for his role in a supposed spying incident involving Siebold, and he was expelled from Nagasaki for portraying its harbor to the dissatisfaction of the Edo regime. Even in the age of photography, painted botanical illustrations continued to have cachet through the works of Takatori Jisuke (b. 1904), especially his *Saishiki shaseizu Nihon no yakuyō shokubutsu* (Color illustrations of Japanese medicinal plants, 2 vols., 1966–1970). To create this compendium, covering 199 species, the artist spent 40 years observing and describing the developmental stages of medicinal herbs, most of them Japanese in origin.[57] Like many of its predecessors, this work is a visual gem, but without the need for geographical comprehensiveness within Japan or the requirement to honor *Bencao gangmu* found in most Edo-period illustrated volumes.

European Visitors Popularize Linnaeus

Examining nature as a specific object, not just a general idea, also found expression in eighteenth-century Japan in a proliferation of travel manuals, tour outlines, crop guides for farmers, elaborately illustrated maps in riotous colors, and Barnum-like lists of odd and remarkable sights. Scholars of herbal medicine and natural historians increasingly collected, described, cataloged, and classified plant specimens as objects of intellectual curiosity as well as items of practical use. In this some were guided by the binomial and sexual classification scheme of Carl Linnaeus, as well as by alternative systems proposed by other European scientists. Linnaeus (1707–1778) transcended the work of various predecessors who sought the natural laws of the plant world, especially in forests, based on visual similarities.[58] Instead, in *Systema Naturae* (1735) he drew on the work of Nehemiah Grew (1641–1712) and Sébastien Vaillant (1669–1722) to propose a system of sexual taxonomy that sorted higher plants into twenty-four classes, depending on numbers of stamens (male sexual structures) and into orders based on numbers of pistils (female sexual structures).[59] Two years later, in *Genera Plantarum,* he described roughly 7,000 plants and established the genus as the primary botanical classifier, and in *Species Plantarum* (1753) he established both genera and species, creating a binomial system of nomenclature for identifying plants. Linnaeus' system, although easy to understand, was artificial and excluded many plant groups, yet his sexual method and binomial terms provoked much research in the later eighteenth century and "provided the building blocks upon which modern botany rested."[60] The French biologist Antoine Laurent de Jussieu (1748–1836) kept Linnaeus' binominal terms but rejected his sexual classification as contrived, instead identifying flowering plants according to multiple characteristics in his *Genera*

Plantarum, published in 1789. If not exactly revolutionary, Jussieu's approach gradually displaced Linnaeus' system among continental botanists, whereas in Great Britain Linnaeus remained the gold standard.[61] In Japan, the Linnaean shift to modern biology slowly displaced the study of natural history and herbal medicine as the Edo period drew to a close.[62]

One point of entry for Linnaeus' ideas to Japan was the botanical research of the Swede Carl Peter Thunberg (1743–1828), a former student of Linnaeus at Uppsala University who served as head surgeon at the Dutch trading post in Nagasaki for fifteen months in 1775–1776. Thunberg met with Japanese physicians, including Nakagawa Jun'an (1739–1786), who became an important translator of Dutch-language medical books into Japanese. Thunberg collected more than 1,000 Japanese plants, mainly while en route to Edo accompanying the head of the Dutch mission in Nagasaki. Against government regulations, he exported to the Netherlands live specimens of Hiba arborvitae (*Thujopsis dolobrata*) and Japanese sago palm (*Cycas revoluta*).[63] The dried herbarium specimens he took home formed the documentation for his *Flora Japonica* (1784), which described 812 plant species, more than 300 of them new to science.[64] This was the first book to use Linnaean taxonomy to classify Japanese plants and provided the most detailed account of Japanese vegetation then available in the West. Thunberg's follow-up volume *Icones Plantarum Japonicarum* (1805) contained fifty plates drawn from the specimens he brought home from Japan. He succeeded Linnaeus at Uppsala and later served as president of the university.[65] Although celebrated in Japan today as a pioneer of Western science in the Edo era, Thunberg enjoyed a mixed scientific reputation and had only a modest impact on Japanese scholars of materia medica and natural history.[66] Still, a number of plant species honor his name, such as beach rose (*Rosa rugosa Thunberg*), and many specialists consider him the "first to bring botany to Japan."[67]

Another popularizer of Linnaeus' binominal terms was the energetic German ophthalmologist Philipp Franz von Siebold (1796–1866), who ministered to the community of twenty Dutch traders as head surgeon at Nagasaki from 1823 to 1829.[68] In 1824 he established a private boarding school, Narutakijuku, at his residence, where he taught Western medicine and also bred numerous Japanese plants. Like Kaempfer 130 years earlier, Siebold was an enthusiastic amateur, not a scientific explorer, who gathered heaps of information about Japanese folklore, culture, and environmental conditions. His cache of material objects was augmented by gifts in kind from Japanese he treated, including the first cataract operation performed in Japan.[69] He interacted with Japanese physicians, mentored a number of young scholars of medicine and herbal remedies, and traveled widely in Japan collecting plants before being expelled from the country in 1829 on suspicion of spying for Tsarist

Russia. His offense was illegally obtaining coastal survey maps; all but one of his twenty-four medical students at Narutakijuku were imprisoned because of the affair, as was the government astronomer who provided him the maps, Takahashi Kageyasu (1785–1829).[70] Siebold shipped the maps and hundreds of Japanese books illicitly to the Netherlands, "at the cost of much suffering by Von Siebold's Japanese friends."[71] As one wry appraisal notes, Siebold's "egoism was often an obstacle to the fulfilment of his ambitions."[72]

Siebold brought 1,200 live plants representing 485 species with him when he returned to Europe; most were confiscated on arrival at Antwerp, but some reached his nursery in Leiden, and others went to a nursery in Ghent.[73] Siebold set about writing eight books on Japan, establishing him as the West's leading authority on that country in the mid-nineteenth century. His *Nippon* (seven parts, 1832–1882), which relied on materials gathered by his pupil Itō Keisuke (1803–1901), was an illustrated ethnography that introduced Japanese archaeology to Europe. Siebold's *Flora Japonica* (1835, 1870), written in collaboration with German botanist Joseph Gerhard Zuccarini (1797–1848), was the most comprehensive study of Japanese plant life to that time and is still considered scientifically accurate, with 150 beautiful illustrations based on Kawahara Keiga's models.[74] Although Siebold taught Linnaean taxonomy to his students with great effect, his *Flora* followed the classifications of Antoine Laurent de Jussieu, as refined by Augustin Pyramus de Candolle (1778–1841), the leading Swiss botanist of his era. *Flora* revised and improved Carl Peter Thunberg's 1784 book by the same title, and it relied also on the work of German naturalist Johann Georg Gmelin the Younger (1709–1755).[75] In these and other publications, Siebold drew extensively on the writings of his predecessors, prompting Federico Marcon to conclude that he was "a rather mediocre naturalist who routinely relied on the skills of other scholars" and on his Japanese students in Nagasaki.[76]

Siebold returned to Japan from 1859 to 1862 as a trade representative of the Dutch government, then was asked to leave by officials in Amsterdam after the Japanese accused him of meddling in treaty port negotiations and having Japanese legal texts translated into Dutch. He departed for the Netherlands with 280 live plant species as well as seeds, fewer than in 1829. In 1863 his catalog for his nursery in Leiden listed 838 species and varieties of Japanese plants for sale; he was clearly responsible for introducing many Japanese ornamentals to Europe (and onward to North America). Among the most popular were climbing hydrangea (*Hydrangea anomala petiolaris*), which grows up trees and rocks to a height of 15 meters with flower clusters as wide as 25 centimeters; Japanese or palmate maple (*Acer palmatum*); and full moon or feather fan maple (*Acer japonicum Thunberg*).[77] Peter Barnes notes that "the great majority of pre-1860 plant introductions from Japan to the West can more or less be attributed to Siebold and his agents."[78] Many Japanese plants in at least eight genera bear Siebold's name, including Siebold's maple (*Acer sieboldianum,* a small tree); Siebold's barberry (*Berberis sieboldii*); clematis florida (*Clematis florida* var. *sieboldiana,* a deciduous vine with showy white flowers); plantain lily (*Hosta sieboldii*); Japanese red elderberry (*Sambucus sieboldii*); and varieties of magnolias, hemlocks, and cherries, all with scientific names ending in *sieboldii.*[79] Siebold's interest in Japanese material culture extended to his son Heinrich von Siebold (1852–1908), who excavated Kanto-area shell middens at Ōmori, Tsurumi, and Ōji in 1877 and later traveled to Hokkaido to study Ainu language, becoming the first to write on Ainu in the West.[80]

Horticulture Spreads throughout Japan

Horticulture, broadly understood as the deliberate growing of plants by humans for ornamentation, developed deep roots in Edo-period Japan partly through the interest shown by Europeans such as Kaempfer, Thunberg, and Siebold. Well beyond the elite circles of shogunal and daimyo-sponsored herbalists were legions of citizen-collectors who thronged flower shows and horticultural exhibits in the later Edo period to enjoy rare plants both visually and by scent. Huge displays of chrysanthemums, first imported from China in the sixth century CE, were staged in Kyoto and Edo starting in 1715, offering as many as 800 varieties in bloom. Carpenters crafted elaborate boxes to ship plants for exhibit. The wide popularity of flower arranging (ikebana), introduced from China in the seventh century and developing distinctively Japanese styles in the fifteenth century, was reflected in the fifty or more books on the subject published later in the Edo era. Networks of amateur cultural clubs for enjoying flowers and plants proliferated side by side with circles for dance, incense, tea, ceramics, and other art forms.[81]

Horticulture may have been a relative latecomer to Japanese aesthetic expression because of the prior establishment of landscape gardening as an art form stimulated by Chinese models as early as 620 CE.[82] Since antiquity cherries, rhododendrons, azaleas, and maples have populated Japan's landscape gardens as elements of an overall design that includes woody plants, water, rocks, earthworks, ground cover, and pathways, often with muted colors for most of the year. Tokugawa Iemitsu (1604–1651, shogun 1623–1651), known for cruelty toward Christians, foreigners, and his own relatives, prized horticulture because he loved the finest flowers.[83] *Kadan kōmoku* (Compendium of flowering plants, 1681) by Mizuno Motokatsu (fl. 1664–1680) is Japan's oldest surviving horticultural manual, including plants from China and a few from Europe introduced by the Dutch. Another noted forerunner was Itō Ihee (1676–1757), a nursery operator at Somei (Komagome) north of Edo castle whose *Kinshū makura* (Brocade pillow, 1692) anticipated the azalea boom that seized Japan for the next half century by describing 171 *tsutsuji* (*Rhododendron dilatatum* or *R. japonicum*) and 161 *satsuki* (*Rhododendron lateritium*) azaleas. By that time all the wild azaleas in Japan's mountains, as well as some from Korea, were cultivated at Japanese nurseries. Interestingly, torch, Kaempfer, or Johanna azalea (*Rhododendron kaempferi*) was widely distributed in Japan as mountain azalea (*yamazakura*) but arrived in the United States only in 1892. *Tsutsuji* are more common around the world today; *satsuki* are late bloomers commonly used for bonsai and landscape design.[84]

By the end of the seventeenth century enthusiasm for horticulture spread well beyond elites to city merchants. The early years of the eighteenth century featured an outpouring of floral guidebooks, manuals, and scholarly studies, and after 1725 the focus of Japanese horticulture shifted from flowering trees and shrubs to ornamental plants such as morning glories, chrysanthemums, primroses, irises, and camellias.[85] By midcentury growing flowers and picking herbs were pastimes for all social classes; the art of tending bonsai gained popularity, and the craft of gardener (*uekiya*) thrived. Although most foreigners were not welcome in the Edo period, their plants certainly were: sunflowers, marigolds, *Mirabilis jalapa* (marvel of Peru), *Canna indica* (a reed known as Indian shot), Southeast Asian palms, hostas from China, and tobacco all entered at this time.[86] Domestic flowers such as irises, peonies, morning glories, and cherries were cultivated with great skill in the eighteenth and nineteenth centuries. Irises, which for many centuries grew wild in mountain wetlands, now enjoyed huge éclat as garden flowers. Japanese horticulturalists planted legions of peonies and battled the ants their blooms attracted. Morning glories underwent two fads among ordinary residents, first in 1804–1830 and then in 1848–1860; because they were annuals, they could be propagated only via seeds. Through genetic variation gardeners developed many varietals well before and quite independent of Mendel's laws, which were announced in 1865 but had little effect before 1900.[87] Japanese mountain cherry (*Prunus jamasakura,* properly *P. serrulata*) was the first wild cherry species to be cultivated in Japanese gardens; by the mid-nineteenth century more than 300 varieties of flowering cherries were found in Japan.[88] Another popular favorite widely seen in Japan today, Japanese wisteria (*Wisteria floribunda*), was bred assiduously in the Edo period in many colors; the stems of this species turn clockwise, not counterclockwise as with wisteria in the West. Because of limited trade, poor transport, and low survival rates during perilous sea voyages, Japan's achievements in horticulture were little recognized in the West before the mid-nineteenth century. When modern botany arrived after 1868, Japan had little to learn from seemingly static European horticulture compared with the steady advancements of Japanese gardeners during the previous two centuries.[89]

Late Edo-Period Plant Scientists

The most elaborately detailed and scientifically accurate work by Japanese naturalists appeared in the final half century of the Edo period. In 1828 Iwasaki Tsunemasa (Kan'en, 1786–1842) produced *Honzō zufu*, an enormous illustrated compendium of more than 1,500 plant species that in its centennial reprinting ran to ninety-six volumes. Although the title contained the word *honzō*, Iwasaki was concerned with describing the plants themselves, not their medicinal uses.[90] Even more forward-looking was the physician and renowned scholar of Western science Udagawa Yōan (1798–1846). Udagawa knew at least four Western languages and translated Dutch medical manuals into Japanese. Much influenced by Siebold's ideas about Western medicine, Udagawa eschewed both herbalism and conventional natural history, instead prefiguring modern botanical science almost before other scholars were ready to consider its approach to the study of plants. In 1822 he wrote *Botanika kyō* (Botanical canon), considered the first Japanese work to use *botanika* for botany and *dōbutsu* for animals. Udagawa explained botany as *shokugaku* (study of plants): "Botany *shokugaku* and materia medica *honzō* are clearly separate sciences."[91] *Shokugaku* was later abandoned in favor of *shokubutsugaku* (both words mean *botany*), a term used in Chinese in 1857 (*zhiwuxue*) and apparently for the first time in Japanese in 1867.[92] His three-volume *Shokugaku keigen* (Basics of botany) was completed between 1833 and 1835, introducing Western plant morphology, physiology, and chemistry in radically new terminology still used today, such as *chisso* (nitrogen), *tanso* (carbon), and *sanso* (oxygen).[93]

Another pioneering botanist was Iinuma Yokusai (1782–1865), a pupil of Ono Ranzan and Siebold. Iinuma served as a doctor until age fifty, then relinquished his practice to plant various species in his garden and study them, as well as pressed flower specimens, through a microscope. In Edo he studied Western science with Udagawa Yōan, an unusual step for someone sixteen years older than his teacher. After two decades' hard work Iinuma produced *Sōmoku zusetsu* (Plant atlas, 20 vols., 1856), which the prolific modern botanist Makino Tomitarō (1862–1957) called the first attempt by a Japanese scholar "to figure and describe our native plants scientifically."[94] In this work Iinuma portrayed 1,201 herbaceous plants according to Linnaeus' system, not the traditional *Bencao gangmu* classifications; he also detailed 600 woody plants in a separate manuscript that was never published. *Sōmoku zusetsu* was considered a big step forward from the work of Iwasaki and Udagawa in thoroughness and precision, advancing Japanese plant studies to the brink of modern science.[95]

Most celebrated and most important of late Edo-period botanists was

Itō Keisuke (1803–1901), who like Udagawa Yōan and many other plant specialists was born to a medical family. He studied with Mizutani Hōbun and later with Siebold in Nagasaki, then as a physician was credited with developing a vaccine against smallpox in 1852. His *Taisei honzō meiso* (Names of Western herbals, 1829) was the first book in Japan to provide Latin scientific as well as Japanese names for the plants included in a copy of Thunberg's *Flora Japonica* given him by Siebold, and Itō's book emulated Thunberg in following Linnaean taxonomy in Western alphabetic order, not the classifications of Chinese herbal medicine. Yet Itō did not so much reject earlier *honzō* scholars, by whom he was much respected, as transcend their work in embracing Western plant science. In 1861 he became a lecturer at the Bansho Shirabesho (Institute for Investigating Foreign Books, est. 1857) in Edo and in 1870 at its successor, the Kaisei Gakkō (Imperial College), which was subsumed in the newly established University of Tokyo in 1877. Although he never attended a university or earned an academic degree, Itō became a visiting professor of botany at the University of Tokyo that year, was promoted to a professorship in botany in 1881, and was named a doctor of science by the university in 1888.[96] In this way he bridged the worlds of herbal medicine and modern science and cleared the track for Japan's scientific revolution that established botany, zoology, and geology as main areas of research on a par with Western counterparts by the end of the nineteenth century.

One emblem of changing times was the conversion of the government herb garden at Koishikawa, with origins tracing to 1638, into the renamed Koishikawa Botanical Gardens (Shokubutsuen), attached to the new University of Tokyo in 1877. Undoubtedly the Meiji restoration of 1868 was "a big turning point in teaching Western botany and zoology apart from *honzōgaku* [study of herbal medicine]."[97] Before then herbalists and naturalists specialized in medicine or Confucianism and had little opportunity to adopt new research paradigms, despite pockets of Western learning (*yōgaku*) in certain Edo-era cities. A few botanical conservatives rued the transition to modern learning: the dendrologist Kitamura Masami (1926–2012) claimed that when Western natural science came to Japan, "it also brought a detachment from nature. Perhaps the greatest loss from this was the feeling of unity with nature."[98] Japan's pluriform scholarship on materia medica and natural history was not a harbinger of modern science, but such Edo-era practices as gathering specimens, conducting surveys, creating detailed images, and describing and classifying plants with terminology still used today were appropriated by the new field of botany starting in the 1870s.[99] Although scholars of herbal medicine resisted the modernization of medicine and pharmacology, Edo-period researchers—operating in parallel with but independently of European counterparts—left a legacy of techniques remarkably similar to the criteria for modern scientific study of plants in the West, as detailed by historian Sara Tovah Scharf: (1) technical vocabulary; (2) order in describing plant parts; (3) point-form notation; (4) layout of text on each page; and (5) indexes and cross-references.[100] Japan began the Edo period with a well-formed perception of plants in imagination and as familiar objects, but by the mid-nineteenth century nature was abundantly classified and illustrated, if not yet well understood through scientific experimentation and analysis. In the Edo era plants were mobile, moving from mountains to landscaped gardens, from country to city, from objects of simple enjoyment to commodities for exhibit and sale, a process that continues to this day.

Three

Seeking Japanese Plants in Europe and North America

In Western Europe a fervid wave of enthusiasm for trees and flowering plants from East Asia began to build in the mid-seventeenth century, surged in the eighteenth and nineteenth centuries, and crested early in the twentieth. To some degree this longing for Chinese, Korean, and Japanese flora stemmed from imperial botany, understood as a quest to bring the world's vegetation home for the delectation of the overseers of empire. Explorers were dispatched around the globe to bring back exotic vegetation to satisfy both curiosity about the foreign and the imperatives of newly emerging life sciences. Plants of culinary or medicinal value to the realm were especially welcome. Networks of colonial gardens in Cape Town, Calcutta (Kolkata), Saint Helena island in the South Atlantic, Batavia (Jakarta), Mauritius, and Saint Vincent island in the Caribbean linked these tropical sites to metropoles in Europe. Lavish illustrations of flora stimulated imperial desire to make overseas colonies more visible at home. At the same time, philosopher Enrique Dussel warns against the "'planetary' perspective of European imperial eyes that became instrumental for bourgeois colonial adventures,"[1] inasmuch as commercial possibilities were never far from mind when plant explorations were being plotted.

Many elements contributed to this European fascination with plants from afar. Early modern Europe was the age of empires, following the voyages of discovery (1500–1800), the Peace of Westphalia (1648) that created sovereign states, and oceanic trade that brought enticingly novel spices, fabrics, and objects from abroad to stir cravings among the wealthy for the exclusive, the beautiful, and the fashionable as commodities for consumption. In the late sixteenth century elite landholders in England began creating private parks as pleasure grounds for themselves and their peers, ornamented with trees and flowers newly imported from around the globe, and they built gardens to grow plants from overseas. A hunger for unseen plants, animals, and artifacts was piqued by accounts from far-flung travelers and reports by Jesuit missionaries abroad, triggering the emergence of collecting as a pastime. The eighteenth-century enlightenment fed an appetite for encyclopedic knowledge and led to a new cosmopolitanism of curiosity about the whole world. In a newly secular age, the rise of scientific inquiry expanded the scope of investigation to the entire planet, revealing that plants knew no borders and were in constant motion.

This chapter outlines the early modern European desire for East Asian flora, then focuses on how both science and commerce spurred plant collecting and shipment to Europe and North America from the early nineteenth century to recent times. Beginning in the late nineteenth century nursery operators catered to the Western delight in plants from Japan with export services from Yokohama to customers around the world. Import restrictions in the United States after 1912 did little to temper the horticultural fascination with ornamentals considered Japanese, even though many traced their origins to ancestors in the eastern Himalaya and western China.

"Plants and Seeds Wanted from China and Japan"

The House of Orange and other wealthy Dutch families sought new plant species from Asia for their pleasure gardens via Dutch East India Company traders (VOC, est. 1602) throughout the seventeenth and eighteenth centuries, as did French estate owners who obtained seeds and specimens sent to Paris by Jesuit missionaries starting in the seventeenth century. Biologist Stephen Spongberg has found that the ginkgo, known to be growing in a botanical garden at Utrecht in the 1730s and at William Hamilton's estate on the Schuylkill river by 1784, was probably the first East Asian tree to be widely cultivated as a shade tree in Europe and North America.[2] Seeds of the common camellia (*Camellia japonica*), considered native to China and Japan, arrived in Europe from China in 1739 and grew well in acidic soil under mild, humid conditions.[3] In 1747 Jesuit priests brought seeds of the pagoda tree (*Sophora japonica L.*), a Chinese medicinal herb called scholar's tree, to Paris, followed in the 1770s by the golden rain tree (*Koelreuteria paniculata laxmann*), an ornamental tree with bright yellow flowers that later was planted at Thomas Jefferson's estate Monticello in 1809. Other French priests founded missions in the interior of China's Yunnan province and collected plants to be sent home by way of Canton (Guangzhou), the only licit port for trade with Europe between 1757 and 1842.[4]

Plants from East Asia were objects of particular fascination in France and England in the second half of the eighteenth century, driven in part by a burgeoning interest in natural history during the age of enlightenment. Scholars in this era amassed their dried herbarium specimens in collections called cabinets.[5] The British Museum owns a 1789 catalog of "plants and seeds wanted from China and Japan," presumably sought by London nursery operators for sale to wealthy European landowners swept up in the prevailing obsession with East Asian greenery. Based partly on Engelbert Kaempfer's research at Nagasaki from 1690 to 1692,[6] the catalog devoted eight pages to varieties of azaleas, camellias, and peonies sought from Japan, and two pages to flowering trees and shrubs from China. The seriousness of this quest can be judged from the elaborate directions provided on how to ship seeds and seedlings from East Asia to Europe, given the great heat of the tropics through which they must pass. Detailed line drawings illustrated the best types of crates to be used, including how to pack plants in wax, moss, or moist soil.[7]

Peonies were known in European gardens since the time of Pliny the Younger (61–113 CE) but had little cachet until the first Chinese peonies arrived in England from Canton in 1789. Within the following decades more than 250 varieties were cultivated in Britain; by the early twentieth century 800 varieties had been imported to the United States from Europe and Japan.[8] Some azaleas from Japan made their way to Europe in the eighteenth and nineteenth centuries, and then from Belgium to the United States. Irises were another favorite in Western Europe, especially the Japanese water iris (*Iris ensata*) with its large, velvety flowers, as was the so-called Japanese anemone (*Anemone hupehensis*), a multicolored, autumn-blooming hybrid of a Chinese and a Japanese species.[9] British enthusiasm for ornamental plants led to the founding of the Horticultural Society of London (later Royal Horticultural Society) in 1804; similar organizations were established at Philadelphia in 1827 and in Boston two years later. The London society soon began receiving chrysanthemums, tree peonies, camellias, azaleas, and other species from an agent of the British East India Company in Canton. By midcentury Chinese ornamentals poured into the Royal Botanic Gardens at Kew, as well as gardens in Edinburgh, Glasgow, and Paris[10]—a hearty and profitable response to the 1789 call for plants and seeds from East Asia.

Philipp Franz von Siebold, recognized as the single most important conduit of Japanese flora to Western Europe in the first half of the nineteenth century, returned from Japan to the Netherlands in July 1830 with specimens of mammals, birds, fish, reptiles, and 5,000 invertebrates, as well as 1,200 live plants and 12,000 dried plant samples. Most of those that survived customs inspection ended up with Siebold in Leiden, where he constructed a house called Nippon with a greenhouse and botanical garden, eventually propagating about 300 kinds of plants from Japanese seeds, none of them visible at his former estate today. Between 1829 and 1844 Japanese gardeners sent Siebold more stock, of which 231 live plants and 204 kinds of seeds arrived safely.[11] These were augmented by 280 live plant species Siebold sent back to Leiden after his second stay in Japan, from 1859 to 1862. Through the nursery he opened at Leiden in 1826, Siebold is credited with introducing to the Netherlands more than 1,000 plants species in 156 genera from Japan, including ginkgos, cherries, chrysanthemums, hydrangeas, lilies, irises, wisteria, camellias, peonies, bamboo, and conifers, most of which were marketed by mail order to horticulturalists throughout Europe and beyond until von Siebold and Company went out of business in 1882. His 1861 nursery catalog listed nearly 4,000 types of plants for sale, mainly from Japan.[12] The botanical garden at Leiden University contains descendants of specimens Siebold transplanted, while the National Herbarium in Leiden has Siebold's collection of 12,000 dried Japanese plant specimens.[13] Today the botanical gardens, public parks, and private landholdings of Western Europe are living laboratories displaying the many plant introductions made by Siebold.

In East Asia the eighteenth- and early nineteenth-century era of overtly imperial botany is thought to have peaked with plant collecting in the Himalaya between 1848 and 1851 led by British botanist and geographer Joseph Dalton Hooker (1817–1911), but private botanical forays on behalf of European and North American botanical gardens, research institutes, and commercial nurseries proliferated into the early twentieth century and have continued occasionally since. The rise of rail and air travel has greatly reduced the need for lengthy treks for onsite investigation and collecting by foreign plant specialists.[14] Once rather imperiously called *explorations* or *expeditions,* such overseas ventures are now more modestly called fieldwork for botanical exchanges with colleagues in host countries.

European Plant Collectors in Japan

After more than two centuries of limited contact with foreigners, Japan signed a Treaty of Amity and Commerce with the United States in 1858, the first of several agreements the Edo government reached with Western powers that granted extraterritorial rights for Westerners to travel, reside, and do business in Japan. For the next two decades a steady inflow of visitors from botanical gardens, research units, and plant retailers in Western Europe and the United States toured Japan in search of unfamiliar trees, shrubs, and flowers for shipment home, as live plants, seeds, and herbarium specimens. The collectors may have thought of themselves as explorers, but they quickly found that Japan was an ancient and cultured civilization, hardly in need of discovery by travelers from abroad. Trips to select plants for export occurred at a slower pace early in the twentieth century and have continued on a more focused, collaborative basis with Japanese specialists ever since.

Among the earliest collectors was John Gould Veitch (1839–1870), scion of the London Chelsea nursery firm James Veitch and Sons (est. 1808) that grew into Europe's most extensive network of family-run plant nurseries in the nineteenth century. Veitch Nurseries employed twenty-two plant hunters between 1840 and 1905 to scour the planet for novel and unusual species before the company went out of business in 1914.[15] John Gould Veitch toured China, Japan, and the Philippines in 1860 and 1861, five months of which were spent in Japan. Veitch arrived at Nagasaki in July 1860, gathered seventeen species of conifers for shipment home as live plants accompanied by seeds, and found time to join Britain's first consul in Japan, Rutherford Alcock (1809–1897), and six other Englishmen in scaling mount Fuji on September 11, 1860—apparently the first non-Japanese to do so. Veitch sketched a verbal map of the successive tiers of vegetation on Fuji, identifying stunted larches (*Abies leptolepis* or *Larix kaempferi*) above 1,800 meters and one or two dwarf conifers above 2,500 meters.[16] In an inventory he published in 1863, Veitch listed 18 varieties of ferns and 63 varieties of plants and trees he sent home from Japan to England, including elms, maples, and conifers.[17]

Veitch is credited with introducing many Japanese plant species to Western botany via the family nursery business, one of which bears his name, Veitch's silver fir (*Abies veitchii*). Others included Japanese zelkova (*Zelkova serrata*), Japanese umbrella pine (*Sciadopitys verticillata*), and several Japanese ferns to cater to a brief Victorian fern craze.[18] He sent home samples of bamboo grass (*Sasa veitchii*) and varieties of bamboo, bulbs of golden-rayed lilies (*Lilium auratum*), Japanese larch (*Larix kaempferi*), northern Japanese hemlock (*Tsuga diversifolia*), and black lily magnolia (*Magnolia liliiflora "nigra"*), with dark red tepals. Tiger lilies (*Lilium lancifolium*), with orange and black blooms, arrived in Europe from Japan at the same time; although their bulbs have been eaten as a vegetable in East Asia for millennia, their pollen is toxic.[19] Perhaps Veitch's most noted introduction to Western botany was Japanese creeper (*Parthenocissus tricuspidata [tsuta]*), often called Boston ivy even though it belongs to the grape family. Japanese creeper, common in the Boston area, can reach eight meters in length and often climbs the walls of Ivy League colleges; Stephen Spongberg considers this plant "the most valuable vine ever introduced into Western horticulture."[20]

John Gould Veitch's rival in gathering plants for export to Great Britain was the Scot Robert Fortune (1822–1880), who first arrived at Suzhou, China, in 1843 to obtain botanical specimens for the Horticultural Society of London. Altogether Fortune made four trips to China between 1843 and 1860 before setting foot in Japan in October 1860. Like other Western collectors at the time, he shipped live woody and herbaceous plants by sea in Wardian cases, sealed glass-walled crates containing moist soil atop wooden bottoms—in effect, minigreenhouses —that almost unfailingly preserved the plants on long voyages through the tropics and temperate zones en route to London: "Nearly every plant of importance has been introduced alive," he confidently stated after the cases reached England.[21] Fortune spent nine months in Japan, mostly avoiding the rigors of countryside and forest for the ease of purchasing plants for shipment back to the U.K. nursery at Bagshot and to the Standish and Noble nursery in Sunningdale, Surrey. Still, he was able to claim in 1862 that "never at any one time had I met with so many fine plants" as those he gathered in Japan.[22] Among the species Fortune introduced (or reintroduced in modern times) to Europe were the male spotted laurel (*Aucuba japonica,* also called Japanese laurel); Sawara cypress (*Chamaecyparis pisifera*); chocolate vine (*Akebia quinata*), considered invasive in the eastern United States; Japanese cryptomeria (*Cryptomeria japonica*), fossils of which date to 7 million years ago in England; winter hazel (*Corylopsis spicata*); Japanese purple meadowsweet (*Filipendula purpurea*); varieties of rhododendron from the northern Honshu mountains; and a little-known woodland species named for Fortune, Japanese rockfoil (*Saxifraga fortunei*), a low, rounded herbaceous perennial with reddish flowers.[23]

Fortune was not shy in generalizing about the Japanese he met during his brief stay. The *Shokubutsugaku no hyakka jiten* (Encyclopedia of botany, 2016) produced by the Botanical Society of Japan notes with asperity that Britain was the source of the conceit that Japan lived in isolation from the world in the Edo period; as a consequence, Fortune exoticized Japanese plants and emphasized their differences from those elsewhere. He also rather patronizingly asserted, "The remarkable characteristic of the Japanese national temper is that everyone, even the lower classes, has an innate love of flowers. For recreation they take supreme pleasure in a little nurturing of their favorite plants. If we take a national character that loves flowers as proof of the height of a people's cultured life, compared with English people of the same class [the Japanese] are far superior."[24] Fortune was of course a product of his era, one in which the U.K. was rapidly rising to first among the great powers, a point not lost on Japan's leaders after the Meiji restoration of 1868 as they looked abroad for Western nations to emulate.

In 1861, the same year Fortune departed Japan for home, botanist William Jackson Hooker (1785–1865) published an inventory of 1,600 Japanese plant species sent live or as seeds by Western collectors, including Veitch and Fortune, home to Britain, where Hooker served as director of the Royal Botanic Gardens, Kew. Hooker's list included ferns, algae, and flowering plants as well as coniferous trees, but provided no commentary or analysis.[25] This already abundant stockpile of Japanese vegetation may explain why Richard Oldham (1837–1864), a gardener at Kew, became the last official plant gatherer sent anywhere overseas by the Royal Botanic Gardens. Oldham arrived in Japan in 1861, spending most of his two years there in the Nagasaki area, then went briefly to Korea before reaching Amoy (Xiamen) shortly before his death in 1864. According to notes compiled in 1867 by botanist Daniel Oliver (1830–1916), librarian of the Kew herbarium, Oldham sent samples of Japanese star magnolia (*Magnolia stellata*) back to London and amassed an extensive herbarium collection of dried Japanese plants.[26] Today the preserved plants Oldham found in East Asia comprise species from 156 genera, the majority from Japan, within the massive Kew herbarium collection of 7 million specimens.[27]

While visitors from Britain were active in Honshu and Kyushu, the Russian botanist Carl Johann Maximowicz (Karl Ivanovich Maksimovich, 1827–1891) arrived at Hakodate, Hokkaido, in September 1860 and traveled throughout Japan for three years collecting numerous bulbs, 300 lots of seeds, and 72 chests containing more than 800 herbarium specimens for shipment home to Saint Petersburg. Assisted by Sugawa Chōnosuke (1841–1925), he also sent 400 live plants back to Russia.[28] Japanese botanists consider Maximowicz' work pioneering for its detailed descriptions of regional flora, especially in southern Ezo (Hokkaido), despite the dangers facing foreigners and the limits on travel outside designated treaty ports at the time.[29] Among the species he introduced to Western science was an azalea known as *Rhododendron macrosepalum*. Maximowicz was the first Western biologist to establish close ties with Japanese scholars of plant science: after returning to Saint Petersburg he corresponded with such leading Japanese botanists as Matsumura Jinzō (1856–1928), Tanaka Yoshio (1838–1916), Tashiro Yasusada (1856–1925), Miyabe Kingo (1860–1951), and Makino Tomitarō (1862–1957), all of whom consulted his expertise and sent him plant samples from time to time. Partly through their efforts, at least twenty-two plant species now bear his name. Between 1877 and 1893 he published *Diagnoses Plantarum Novarum Asiaticum* in seven parts, a work that relied extensively on Iinuma Yokusai's illustrated *Sōmoku zusetsu* (Plant atlas, 20 vols., 1856).[30]

Bamboo was popularized as a pesky garden plant by the diplomat Algernon Freeman-Mitford (1837–1916), who served as second consul in the British legation in Edo and Tokyo from 1866 to 1870. Returning to England in 1870, he established a large arboretum with a variety of bamboos capable of withstanding the Gloucestershire winters.[31] Mitford is best known for his *Tales of Old Japan* (1871), which helped spread knowledge of the kabuki play *Kanadehon chūshingura* (Tale of the forty-seven rōnin, 1748) and led to his serving as adviser to Gilbert and Sullivan for their 1885 comic opera *The Mikado.* Mitford published *The Bamboo Garden* in 1896, with detailed descriptions of species and instructions on proper soil and siting for hardy bamboos to ornament English estates and arboretums.[32]

Like some of the plant enthusiasts at the Dutch enclave in Nagasaki, the French native Paul Amadée Ludovic Savatier (1830–1891) was a medical doctor posted to Japan who was fascinated by Japanese flora. An officer in the French navy, he arrived at Yokosuka in 1866 as part of an effort to build a Japanese naval force and spent all but two years of the following decade ministering to workers at the shipyard foundry there. Collaborating with Japanese botanists Itō Keisuke and Tanaka Yoshio, Savatier amassed a large collection of preserved plant specimens that are now housed at the Museum of Natural History in Paris. He and the botanist Adrien René Franchet, based at the museum, compiled *Enumeratio Plantarum in Japonia Sponte Crescentium* (List of plants naturally growing in Japan, 2 vols., 1875–1879), which summarized earlier work by Philipp Franz von Siebold, Asa Gray, Carl Johann Maximowicz, and Dutch botanist Friedrich Anton Wilhelm Miquel (1830–1891) and identified 2,743 seed-bearing plants and 198 ferns.[33] As much as any other book on Japan by foreign botanists produced in the nineteenth century, this work earned high accolades from plant specialists in Japan.[34] Writing in 1874 for the international community in Japan, Savatier credited "the persevering investigations of Japanese botanists, who pursue the science with ardour, and have a far larger acquaintance with the vegetation of this country than is generally supposed in Europe." Savatier praised the Japanese for "their rich herbaries, and their innumerable collections of illustrations, where are exhibited, often in great perfection and with excellent regard to anatomical floral details, almost all the Japanese species diffused among our own herbaries, together with many others still unknown to us."[35] Doubtless because he lived in Japan much longer than Robert Fortune, Savatier's outlook on his Japanese hosts was far less imperious and far more nuanced than Fortune's hasty generalizations about the Japanese "national temper," noted above.

As in China, French missionaries in Japan were sometimes avid plant hunters. Urbain Jean Faurie (1847–1914) arrived in 1874 and lived in Niigata, Hakodate, northern and central Honshu, and Kyushu for the next twenty-three years. Among the most energetic Western plant collectors in Japan, he assembled 22,500 dried specimens for shipment to Paris and was regarded as more knowledgeable about Japanese botany than any other foreigner. His lichen specimens are especially well known; his name is attached to the scientific terms for various genera of Japanese alpine poppies, rhododendron, ferns, and pears.[36]

The Veitch nurseries sent botanist Charles Maries (1851–1902) to China, Taiwan, and Japan from 1877 to 1879 to collect hardy plants for the British market. In Japan alone he gathered 500 live plants for shipment to London, including hydrangeas, lilies, maples, conifers, and magnolias. He was the first foreign biologist to reconnoiter the forests of Hokkaido, which

Figure 3.1 Europeans and Americans were charmed by Japanese landscape architecture at the turn of the twentieth century. Rikugien (1695) is a former daimyo garden donated to Tokyo by the Iwasaki family and opened as a public park in 1938.

were relatively underexplored before Japanese colonization of the island in the 1870s. There he discovered the Maries fir (*Abies mariesii*), one of several plants new to science bearing his name, and he is credited with introducing dozens of species and varieties to Britain. Among these are the variegated-leaf hardy kiwi (*Acinidia kolomikta*), a ten-meter-long vine that survives winter temperatures as cold as minus 40 degrees Celsius, and the Nikko maple (*Acer maximowiczianum*), identified by Maries in a Hokkaido forest in 1881, a tree well known to specialists for its unconventional trifoliate leaves but seldom seen outside arboretums.[37] A mature example lives in the Arnold Arboretum in Jamaica Plain, Massachusetts. Both Faurie and Maries are best known for the plant collections they accumulated rather than for their writings.

The stream of European plant hunters in Japan slacked off after 1880, but well into the next century plants kept arriving in Britain, France, the Netherlands, and elsewhere because of demand from horticulturalists and family gardeners. A bonsai boom began with the first show outside Japan at the Universal Exposition at Paris in 1878, followed by an exhibit of 600 plants in 1904 in New York and a similar event at London in 1909.[38] As a young man Collingwood Ingram (1880–1981) introduced the great white cherry (*Prunus taihaku*) to Britain in 1900. By 1932 he had become a respected authority on cherries and was called on to reintroduce this species to Japan, where it had become extinct.[39] A vogue for so-called Japanese gardens seized Britain during the first decade of the twentieth century, and cherries were especially prized. The Veitch nurseries imported *fugenzo* cherries (*Prunus fugenzo*), with profuse pink double flowers; Cheal's weeping cherries (*Prunus "kiku shidare zakura"*), with clustered pink double flowers; and Kanzan flowering cherries (*Prunus kanzan*), featuring pink double blossoms that bloom before leaves appear. Kanzan cherries arrived in Europe in 1910 and have become the most frequently installed variety of Japanese cherry. Magnolias, maples, azaleas, and wisteria from Japan have had an impact similar to cherries. After World War Two Japanese hostas (especially the large *Hosta sieboldiana*) came into fashion and were widely used in European gardens. Species of orange day lilies from Japan in the genus *Hemerocallis* similarly gained favor after 1945, although today most day lilies bred in Europe or North America have "no direct connection with Japan."[40] These and many other popular species in European horticulture stem from remote ancestors in the Himalaya and were regarded as Japanese plants a generation or more ago, but today they are so thoroughly indigenized in Europe that they are seldom considered foreign imports any longer.

Japanese Plants Arrive in America

Trees, shrubs, and flowers from East Asia enjoyed much the same popularity among the wealthy along the American eastern seaboard in the early nineteenth century as was true in Western Europe toward the end of the eighteenth. For many years horticulturalists in the United States took their cue from counterparts in Great Britain; even during the American War of Independence zealous gardeners traded seeds and plants across the Atlantic.[41] William Hamilton (1745–1813) inherited 300 acres along the west bank of Philadelphia's Schuylkill river and built an extensive collection of flora surrounding his home, Woodlands, where he planted many species from East Asia, imported via Europe. Hamilton introduced to North America the ginkgo, tree of heaven (*Ailanthus altissima*), and paper mulberry (*Broussonetia papyrifera*), among others.[42] Like Philadelphia, Boston was another early hothouse of enthusiasm for plants from abroad: as Sara Butler points out, "Horticulture appealed to early-nineteenth-century Boston elites as a cultural symbol of status and sophistication."[43] In 1801 Harvard University established a botanical garden along the Charles river in Cambridge, but as of 1818 its collection included only a handful of plants from East Asia.[44] A significant infusion of plants from Japan occurred in the 1840s, when species Philipp Franz von Siebold introduced to Europe in 1829 began to reach gardeners and institutions in the United States.

Agronomist James H. Morrow (1820–1865) was the first American to exchange plant seeds with representatives of the Edo government and, together with China-based missionary Samuel Wells Williams (1812–1884), the first to send seeds and herbarium specimens from Japan to the United States. Assigned by Secretary of State Edward Everett (1794–1865) to the Perry expedition to Japan in 1853 and 1854, Morrow's remit was to gather seeds from every country visited and "also to collect plants to bring home & send home & dried specimens & description of plants & their mode of cultivating them."[45] He received a bag from the shogun with 39 kinds of seeds, mainly from vegetables. To these he added other seeds he gathered near Yokohama, noting in his diary for April 3, 1854, that he spent "all day putting up seed to go home by the Saratoga," a naval vessel. "I put in small papers more than two hundred papers of seed"[46] and shipped them to gardeners in New York, Philadelphia, and South Carolina. On the return voyage by way of China, Morrow brought 17 cases of dried and living plants, dispatching herbarium specimens he had pressed by hand to Asa Gray at Harvard, who despite their moldy condition identified 41 new species and 1 new genus among them.[47] The live plants, if they survived, were unaccounted for.

The first direct shipment of live plants from Japan to the United States took place in 1861 from the Yokohama garden of George Rogers Hall (1820–1899), a medical doctor turned horticulturalist who lived in Yokohama between 1859 and 1862. Hall cultivated many species, accepted others from Philipp Franz von Siebold during Siebold's second sojourn in Japan (1859–1862), and apparently also shared collections in Yokohama with Robert Fortune. Hall sent examples of eighteen trees as seedlings in Wardian cases intended for Francis L. Lee (1830–1922) in Chestnut Hill, Massachusetts. By the end of the seventy-day ocean journey, Lee had joined the Union army and consigned the cargo to historian and avid gardener Francis Parkman, Jr. (1823–1893), whose Brookline, Massachusetts, summerhouse bordered

Jamaica Pond. Parkman planted the specimens on his estate and showered them with meticulous care. A second shipment in 1862 accompanied Hall on his return trip home, after which he personally took seeds and six cases of live plants to Parsons and Company, nursery operators in Flushing, New York.[48]

Hall's two shipments telescoped into less than a year the direct import to North America of many Japanese species that took decades to enter Europe beginning with Siebold in 1829. Botanists credit him with introducing Kobus magnolias (*Magnolia kobus*), star magnolias (*Magnolia stellata*), ten forms of sawara cypress (*Chamaecyparis pisifera*), hinoki cypress (*Chamaecyparis obtusa*), and the elm-like Japanese zelkova (*Zelkova serrata*). Hall also introduced several forms of Japanese maples (*Acer palmatum Thunberg*), Japanese wisteria (*Wisteria floribunda [willdenow] de candolle*), Fuji rhododendrons (*Rhododendron brachycarpum* G. Don., now an endangered species), umbrella pines (*Sciadopitys verticillata [thunberg] siebold* and *zuccarini*) with leaves arranged like spokes, and Japanese honeysuckle (*Lonicera japonica* var. *halliana*),[49] the latter considered invasive by many specialists who voiced reproof by attaching Hall's name to the variety he brought to the eastern United States.

In 1872 Hall inherited his father's estate on Narragansett bay in Bristol, Rhode Island, and planted it with many Japanese species, including zelkovas and a hornbeam that still survive. Charles Sprague Sargent (1841–1927), director of Boston's Arnold Arboretum, visited Hall's estate in 1889 and enthused, "Hall's Japanese plants . . . have grown into what, I believe, are the finest specimens of certain species which can be found in the eastern states. Some of them I have not seen equaled in Europe."[50] Best known may be his Japanese yews (*Taxus cuspidata*), the first in the United States, several of which he planted in a courtyard behind his house. The largest yew earned renewed praise in 1916 from Sargent, at which point it supposedly measured 37 meters both in height and circumference. This tree died in 1927; another yew, planted nearby after Hall died in 1899, now spreads nearly 40 meters in circumference and is sometimes mistaken for the original Hall brought from Japan in 1862.[51] For much of the twentieth century the Japanese yew was "the quintessential landscape shrub in northern cities and towns" in the United States,[52] although it has lost stature with many horticulturalists today. Less well known in North America than Siebold is in Europe, Hall nonetheless ranks as the first and arguably most important conduit of plant life from Japan to the western hemisphere in the nineteenth century.

Another medical doctor turned horticulturalist was Thomas Hogg, Jr. (1820–1892), an exact contemporary of Hall's who was appointed by President Abraham Lincoln as a U.S. marshal posted to Kanagawa (Yokohama) in 1862. Born in London as the son of a Scottish nursery operator, Hogg arrived in the United States as an infant and eventually became a neurosurgeon in New York.

Figure 3.2 Giant Japanese yew planted shortly after 1899 on estate of George Rogers Hall, M.D., in Bristol, Rhode Island, 2017. Courtesy Karen L. Thornber.

After moving to Japan he established a garden, met the Russian botanist Carl Johann Maximowicz, and rose to the position of United States consul before returning to New York in 1869. Hogg went back to Japan in 1874 and 1875 at the invitation of the Japanese government (*yatoi gaikokujin*) to advise its customs service. During his first period in Japan he shipped plants and seeds, mostly the first of their kind in North America, to his family's nursery business in midtown Manhattan and to his brother James' garden at 84th Street and the East river in Yorkville. The family had closed the business by the time of Hogg's return to Japan in 1874, so he sent the seeds he collected from 150 kinds of trees and shrubs to Parsons and Company in Flushing.[53]

Like Hall, Hogg sent various plants from Japan directly to the United

States that Siebold, Fortune, Maximowicz, and John Gould Veitch had earlier brought to Europe and were heretofore available in North America only indirectly by import across the Atlantic. Peter Del Tredeci has identified 40 hardy, woody species and 14 documented cultivars introduced by Hogg,[54] including deciduous camellia (*Stewartia pseudocamellia*), Japanese snowbell (*Styrax japonica*), fragrant snowbell (*Styrax obassia*) from Hokkaido, and the katsura tree (*Cercidiphyllum japonicum*), all commonly used in North American landscapes. He also introduced several types of magnolias, hostas (*Hosta decorata*), Japanese water iris (*Iris ensata*), and sapphire berries (*Simplocos paniculata (thunb.) miq.*, also called Asian sweetleaf).[55] Also like Hall, Hogg is notorious among ecologists for introducing via the Parsons nursery a plant considered invasive, kudzu (*Pueraria montana* var. *lobata, Pueraria lobata (willd.) ohwi*), the high-climbing vine that escaped cultivation in the southeast United States and often crowds out other vegetation. Japanese knotweed (*Reynoutria japonica* or *Fallopia japonica*) came to North America in the 1860s by way of Europe, introduced there earlier by Siebold, and Hogg sent more of it in 1867 or 1868. Hogg is also thought to have introduced oriental bittersweet (*Celastrus orbiculatus*), considered invasive in the eastern United States for its thin, woody vines that encircle and sometimes strangle host trees or shrubs.[56] Nonetheless Hogg stands shoulder to shoulder with Hall as a key source of the ornamental vegetation from Japan that now enhances North American landscapes.

First initiated by the Perry expedition in 1853–1854, the exchange of seeds and plants continued during the Meiji period (1868–1912), including the arrival in Japan of London plane trees (*Plantanus* x *hispanica, Platanus* x *acerifolia*) and American tulip trees (*Liriodendrum tulipifera*). Considerably later, Western cultivars of such Japanese plants as maples, hostas, and hydrangeas came to Japan in the twentieth century. On the advice of former U.S. Commissioner of Agriculture Horace Capron (1804–1885), in 1873 the Hokkaido Development Commission (Kaitakushi, 1870–1881) planted apple trees from abroad, including the United States, around a wooden government office building in Sapporo. By that year 2 million fruit trees and grape vines were distributed throughout Japan, mainly from overseas.[57] More than a century later, cuttings of the 3,000 Japanese cherries around the Tidal Basin in Washington, D.C., donated by Tokyo Mayor Ozaki Yukio in 1912 were given to horticulturalists from Japan in 1980 to replace hundreds of trees lost in a flood when the course of the Ara river in Tokyo shifted, enabling a fresh row of flowering cherries to be planted along its banks.[58]

Among the most famous foreign employees of the Meiji government was William Smith Clark (1826–1886), president of Massachusetts Agricultural College (now University of Massachusetts Amherst), who came with colleagues to Japan for nine months beginning in July 1876 to establish Sapporo Agricultural College (now Hokkaido University) and create an experiment station in that city where American farm techniques were taught. In addition to collecting 166 vascular plant specimens and 43 lichen samples for the Massachusetts college's herbarium, Clark and his associates sent 79 vascular plant samples and numerous seeds from trees and shrubs to Amherst. In 1877 the Massachusetts Agricultural College nursery began growing 3,000 umbrella pines, 200 katsuras, 23 Japanese maples, and various evergreens from these seeds; a dozen species sent from Sapporo to Amherst are currently growing in that town, including some on the UMass campus.[59] Clark also sent Japanese plant seeds to the Arnold Arboretum, including seven species new to that fledgling facility, founded in 1872. Among the most important were katsuras, Kobus magnolias, and Japanese climbing hydrangeas (*Schizophragma hydrangeoides*), the latter a wall-climbing vine that grows to a height of nine meters.[60] Clark's heritage lives not only in Japan, where his parting words to students in 1877—"Boys, be ambitious" (*shōnen yo taishi o idake*)—are known to schoolchildren throughout the country today, but also in North America through his dissemination of Japanese plant species to Massachusetts and eventually far beyond.

Plants and Commerce

The plants from Japan sent home to Europe and America by nineteenth-century Westerners became widely distributed to horticulturalists and landscape gardeners chiefly through commercial nursery operators such as Philipp Franz von Siebold in Leiden; James Veitch and Sons in London; Robert Wallace and Company and New Plant and Bulb Company, both in Colchester, England; and Wallace's rival and future partner Barr and Sons, which eventually merged as Wallace and Barr. In the United States the Hogg family nursery on Broadway at 23rd Street in New York and Parsons and Company on Long Island were soon joined by S. M. Japanese Nursery Company in West Orange, New Jersey, and other profitable importers as interest in Japanese ornamentals swelled by the end of the century.[61]

In Japan several businesses arose during the Meiji period to service the demand from abroad. A pioneer was the German botanist Carl Kramer (1844–?), who first visited Japan from 1867 to 1871 on behalf of the Veitch nurseries, then from 1872 to 1879 served as an occasional adviser to the Japanese government on botany. The archives of the Royal Botanic Gardens, Kew, contain a letter dated December 31, 1870, from Kramer to botanist Joseph Dalton Hooker in London, explaining that illness had delayed his shipping Japanese lilies and seeds as requested by Hooker.[62] More than a year later he set up a small nursery on the Bluff in Yokohama and publicized his business with a classified ad in the February 26, 1872, *London and China Telegraph*:[63]

> C. Kramer begs to inform the public that he has established himself as NURSERYMAN and SEEDSMAN, and is fully prepared to execute any orders for Japanese plants, bulbs, or seeds for any part of the world.
>
> An experience of four years in this business in foreign parts, two of which has been gained in this country[,] may insure the confidence of customers.
>
> Orders must be accompanied by remittances or references, and letters must be addressed to
>
> C. KRAMER
> Care of W. H. Smith, United Club
> Yokohama

Kramer closed the nursery in 1874, briefly taught botany in Tokyo, then moved to Kagoshima and Nagasaki for three years as an adviser on botany and medicinal plants.[64]

Far more influential was L. Boehmer and Company, the only nursery operated by a Westerner in Japan between 1882 and 1908. Louis Boehmer (1843–1896), a Prussian-born horticulturalist, was chosen by U.S. President Ulysses S. Grant as a member of the Capron mission to advise the Meiji state on developing Hokkaido. Boehmer arrived in 1872 with apples, cherries, grapes, and seeds of ornamentals, which were planted in an experimental garden in Aoyama, Tokyo. Boehmer served in Hokkaido from 1874 to 1882, built a greenhouse that became part of the Sapporo Agriculture College botanic garden, and wrote a report for the Hokkaido Development Commission titled *Report of a Botanical Journey in Yesso, in 1874*. Here he noted that he had gathered 120 different types of living plants and 500 dried specimens during six months of collecting in Hokkaido. "Most remarkable were the Azaleas and Dievillias [diervillas] which were growing in great profusion on Hakodate Head and with their pink and scarlet flowers produced a magnificent effect."[65] Boehmer taught farmers how to grow hops, vegetables, apples, pears, and grapes, then stopped working for the Development Commission in 1882 to found L. Boehmer and Company on the Bluff in Yokohama.[66]

There he worked with Alfred Unger (1865–1938) and a Japanese supplier to develop a brisk export business specializing in Japanese trees and ornamental plants until he fell ill and returned to Germany in 1892, where he died four years later. Boehmer was the first to cultivate many species from Hokkaido, such as wheel lily (*Lilium medeoloides*), with vivid orange hanging flowers (racemes); bluebird lacecap hydrangea (*Hydrangea serrata* var. *acuminate*);

climbing hydrangea (*Schizophragma hydrangeoides*); Japanese tree lilac (*Syringa reticulata*); and deciduous Japanese azalea (*Rhododendron japonicum*).[67] L. Boehmer and Company became a vital pathway for knowledge of Japanese horticulture to the rest of the world, especially via the meticulously illustrated catalogs it published under Unger's leadership between 1893 and 1908, when Unger closed the business and returned to Germany. The 1903 catalog, for example, stated that the company annually exported several million "tender lilybulbs," as well as seeds and plants by fast steamers; because the route by way of Suez was long and fraught with temperature changes, sometimes Unger sent material to Europe via the United States. Starting in 1902 some lily bulbs were dispatched in refrigerated cases.[68] With just slight overstatement, the catalog claimed that "amateurs and nurserymen alike have long recognized the value of Japanese plants and we now find them generally distributed in Europe, in North America and in many parts of Australia, even to the neglect of similar native plants. . . . Moreover have the rational methods of western gardeners so improved the original, imported stock and so freely propagated and popularized many kinds, that they are now more common in the garden abroad than in Japan."[69] John L. Creech later summarized the impact of Boehmer's and Unger's efforts: "During the life of the Boehmer Company probably more ornamental plants entered western gardens from Japan than from any other source."[70]

Boehmer's staff was initially headed by Suzuki Uhei (1834–1910) and his son Suzuki Hamakichi (d. 1925), who in 1891 branched off to found the first Japanese-owned nursery for export, initially called Yokohama Gardeners Association but soon changed to Yokohama Nursery Company. This firm immediately established a branch in San Francisco and quickly outstripped L. Boehmer and Company in patronage; it remains in business today, with branches in Hokkaido and Kyushu. When Unger liquidated the Boehmer concern in 1908, its greenhouses were dismantled and sold to Yokohama Nursery Company.[71] Like L. Boehmer and Company, starting in 1901 Yokohama Nursery Company annually issued beautifully illustrated catalogs. The first listed 51 varieties of ornamentals, as well as trees, ferns, aquatic plants, bonsais, seeds, bulbs, and economic plants, with instructions on planting and care. Subsequent catalogs often exceeded one hundred pages, reaching a high point of 118 pages before wartime stringencies took effect in 1937.[72] By then the company had expanded abroad with branches in New York, London, and Shanghai. With their detailed descriptions and fine illustrations, these annual catalogs, together with those from the Boehmer firm, constitute invaluable sources for understanding horticultural relations between Japan and the outside world from the late nineteenth century to World War Two.

Plant Exports to the United States since the Late Nineteenth Century

The global fascination with plants from East Asia that led to brisk botanical commerce also spurred the creation of arboretums and botanical gardens in the later nineteenth and early twentieth centuries. Henry Shaw (1800–1889), a British-born philanthropist, established the first great garden in America, the Missouri Botanical Garden, at Saint Louis in 1859. The Arnold Arboretum, America's first major arboretum, was founded in 1872 on farmland in the Jamaica Plain area of Boston. Both institutions quickly became leading centers of botanical research and practice, joined by the University of California Botanical Garden at Berkeley (1890); the New York Botanical Garden (1891); Longwood Gardens in Kennett Square, Pennsylvania (1906); the National Arboretum in Washington, D.C. (1927); and the Connecticut College Arboretum (1931). Nursery operators in Boston, New York, Philadelphia, and elsewhere taught horticulture to home gardeners on private estates.[73] *The American Gardener* published an eight-part series in 1891 introducing the economic plants of Japan, pointing out the ubiquity of mandarin oranges, grapes, pears, and chestnuts and concluding that the persimmon "is appreciated by all; it is *the* fruit of Japan"[74]—still true in late autumn today. So great was the demand for Japanese ornamentals that the venerable Osaka art dealer Yamanaka and Company, which established a branch near New York's Madison Square in 1894, held a four-day auction of 430 "rare and beautiful Japanese and arboreal plants" at its new Boston branch in November 1899.[75]

Europeans and Americans kept energetically hunting plants in Japan for export from the 1890s until the United States imposed restrictions on importing live plants to curb pests, diseases, and viruses beginning in 1912 and 1919. Charles Sprague Sargent, director of the Arnold Arboretum, teamed with James H. Veitch (1868–1907) of the London nursery firm in 1892 to collect seeds and herbarium specimens in northeastern Japan. James Veitch was two years old when his father, John Gould Veitch, died of tuberculosis. Like his father, James climbed mount Fuji, staying overnight in a hut in August 1892 with ten other men who "presumably slept, but insects engaged the major portion of my attention."[76] At an inn in nearby Enoshima, his travelogue reported, the "quilts are very warm and comfortable, but needless to say, the animal kingdom is largely represented."[77] He generally liked the Koishikawa Botanical Gardens in Tokyo, although he complained that "the herbaceous grounds and shrubbery are very full of labels but little else," and the acacias and weeping willows "are most cruelly pruned, the heads being cut till they are flat."[78]

For his part, Sargent collected several hundred plants and botanical specimens in Hokkaido, guided by the advice of Miyabe Kingo (1860–1951), a professor of botany at Sapporo Agricultural College who in following years supplied Sargent with numerous seeds. Sargent and Veitch also gathered seeds from plants along the shore of lake Chūzenji at Nikkō and at the Hakkōda mountains in northeastern Honshu. Sargent sailed for the United States in November 1892 with seeds of 200 plant species and 1,225 herbarium specimens. Perhaps his most important introduction to North America was the hardy evergreen azalea known as torch, Kaempfer, or Johanna azalea (*Rhododendron kaempferi*), with its profuse red or pink flowers. He also brought back Nikko maples (*Acer maximowiczianum*), Sargent crabapples (*Malus sargentii*), Sargent cherries (*Prunus sargentii*), and longstalk holly (*Ilex pedunculosa miquel*).[79] His travels in Hokkaido and Honshu formed the basis for Sargent's *Forest Flora of Japan* (1894), in which he wrote that "the superiority of the forests of Japan in the number of their arborescent species over those of every other temperate region, eastern North America included, in proportion to their area, has never been fully stated" before. "The proportion of trees to the whole flora of Japan is remarkable," he recorded.[80]

The first two decades of the twentieth century were the grand era of plant collecting by Westerners in East Asia, with less attention given to Japan after 1910 because by then Japanese trees and flowers were so well known abroad. The Arnold Arboretum took the lead in introducing ornamentals to the United States, while the U.S. Department of Agriculture focused mainly but not exclusively on economic crops from 1898 until 1956, when it started partnering with Longwood Gardens to import ornamental plants from around the globe. Altogether the U.S.D.A. has sponsored more than 150 plant-collecting trips to dozens of countries. David Fairchild (1869–1954), who for many years headed the plant introduction section at the U.S.D.A., was an earnest competitor of Sargent's, yet the Arnold Arboretum and the Department of Agriculture apparently cooperated smoothly.[81] Fairchild visited Japan in 1902, where he gathered seeds of various agricultural crops, including Japanese lawngrass (*Zoysia japonica*), a number of ornamental plants, 18 species of bamboo, and 30 varieties of flowering cherries. The cherries died in transit, so five years later Fairchild successfully installed 300 cherry seedlings from Yokohama Nursery Company for Chevy Chase, Maryland, and 150 weeping cherries for Washington, D.C.,[82] forerunners of the 3,000 flowering cherries donated by Mayor Ozaki Yukio of Tokyo to the American capital in 1912. Fairchild's writings show that he was especially proud of introducing wasabi (*Eutrema japonica*), sometimes called Japanese horseradish; Japanese spikenard (*Aralia cordata*), the shoots of which are served sliced, "crisper than celery"; and Japanese paperbush (*Edgeworthia papyrifera*), the long fibers of which are used to make traditional Japanese paper (*washi*) and banknotes.[83] Of these, only wasabi came into widespread use in the United States.

The Arnold Arboretum sent dendrologist John George Jack (1861–1949), a lecturer on botany there, to Japan, then Korea and China, to collect further plant specimens in 1905. In Sapporo he met botanist Miyabe Kingo, Sargent's acquaintance since 1892. From Japan he brought seeds from the Japanese beautyberry (*Callicarpa japonica*), which makes purple drupes (berries) in the fall, inedible by humans but admired for their beauty. Altogether Jack amassed more than 600 lots of seeds and live plants in East Asia for the arboretum.[84] Southeastern Brookline, Massachusetts, near Jamaica Pond and just two kilometers from the Arnold Arboretum, was an important locus of horticulture in the nineteenth and early twentieth centuries, including gardens surrounding the residences of Parkman, Sargent, and Frederick Law Olmsted (1822–1903). Nearby was the estate of Larz Anderson (1866–1937), an American diplomat who first visited Japan in 1889 on a grand tour after graduating from Harvard the previous year. He returned to the United States with two Japanese dwarf maples, married Isabel Weld Perkins (1876–1948), and in 1898 inherited a twenty-six-hectare property from the Perkins family where Anderson built a large garden in 1907. Five years later he was named American ambassador to Japan but served for only one day in March 1913 before the Woodrow Wilson administration came to office in Washington, D.C. That spring Anderson bought more than forty bonsai plants from Yokohama Nursery Company, the oldest of which dated to 1737, and brought them back to his Brookline residence. His collection, one of the springboards of America's bonsai boom, later was moved to the Arnold Arboretum in stages, but by 1989 just fifteen plants survived.[85]

Perhaps the greatest Western plant collector in East Asia between 1900 and 1920 was Ernest Henry Wilson (1876–1930), both because of his work for Veitch and Company and because of the trove he brought back to the Arnold Arboretum after several visits to China. Originally a representative of the Veitch nurseries in London, Wilson was hired by Sargent in 1907 to obtain plants from Hubei and Sichuan, China. From this and other trips, he is thought to have introduced nearly 2,000 plant species (mostly from China) to the West, 382 of them new to science, and prepared 50,000 herbarium specimens.[86] Sargent also sent Wilson to Japan in 1914 to document conifers and Japan's indigenous species of cherries; Wilson photographed every known Japanese conifer in situ and prepared herbarium specimens of eight tree and shrub species, including sixty-seven varieties of cherries.[87] One result of this trip was Wilson's *Cherries of Japan* (1916), which explained morphological distinctions among species and pointed out that five of them produced double flowers, in contrast with those in Europe and North America.[88] Sargent dispatched him again from 1917 to 1919 to Japan, Korea, and China, during which Wilson introduced the popular Kurume azalea (*Rhododendron kurume*) from Kyushu to the

United States.[89] For unbounded energy, resilience in overcoming a broken leg in China in 1910, and distance traveled in search of new specimens, Wilson was unmatched among the plant collectors of his generation.

Another collector of note was Frank Meyer (1875–1918), who made four trips for the U.S.D.A. to China between 1905 and his death there in 1918. Meyer stopped in Japan for eleven days in 1915 and found that chestnut blight was widespread but unrecognized there, suggesting that many Japanese chestnuts were infected by the blight fungus but not killed by it. Trees imported from Japan to North America could thus carry the fungus unnoticed.[90] In 1919 the U.S. Department of Agriculture issued Regulation 37, which restricted imports of live plants from overseas unless first quarantined in U.S.D.A. facilities and inspected for diseases and pests. Henceforth virtually all overseas gathering of living specimens was done through the U.S.D.A., which launched nearly forty collecting teams worldwide during the next two decades. For example, Rolla Kent Beattie (1875–1960), a plant pathologist for the Department of Agriculture, spent the years 1927 through 1930 in Japan and brought back seeds of Japanese chestnuts (*Castanea crenata*) to replace the diseased American species, as well as ninety varieties of azaleas for breeding at U.S.D.A. facilities.[91] As late as the 1970s the import of cherries to the United States was regulated to preclude viruses; clone material was quarantined, but seeds could be freely brought in. These restrictions, together with the rise of air travel, modern communications, and the use of camera film and plastics for collecting, meant that the nature of gathering plants overseas was radically transformed in the latter half of the twentieth century.[92]

During the American occupation of Japan (1945–1952) the U.S. Department of Agriculture carried out research on Japanese grain products with a view toward increasing crop yields in the global south. Meanwhile the huge expansion of residential suburbs in the United States after World War Two meant lively business for horticulturalists and landscapers to satisfy homeowners' demand for handsome plantings. In response, the joint U.S.D.A.-Longwood Garden ornamental plant exploration program, inaugurated in 1956 and partly funded by the Longwood endowment, used U.S.D.A. staff and facilities to organize thirteen plant-gathering forays between 1956 and 1971, nine of them to countries bordering China. John L. Creech (1920–2009), who later directed the National Arboretum in Washington, D.C., also led several trips to Japan seeking plants to introduce to the American east coast. Creech and his team visited evergreen forests in Kyushu, Shikoku, and western Honshu, returning home with samples of Japanese cryptomeria (*Cryptomeria japonica*), hinoki cypress (*Chamaecyparis obtusa*), and Japanese crepe myrtle trees (*Lagerstroemia fauriei*), as well as many other forest plants.[93] Hirado azaleas (*Rhododendron* x *pulchrum*), evergreen hybrids cultivated in Japan since at least the early seventeenth century, were finally brought to the United States in 1961. Meanwhile, in response to the growing market in the United States and elsewhere, more than 80 percent of Japan's plant exports as of 1970 came from Kanagawa prefecture, including Yokohama and the prefectural botanical garden at Ōfuna. By that date the prefectural garden claimed to have developed 900 new varieties of peonies and to export each year 120 varieties of irises, 50 different kinds of lilies, and 170 kinds of azaleas, with total overseas sales of 915,000 plants.[94] The Ōfuna facility continues to service overseas markets today.

The Arnold Arboretum sent seed collectors to Japan and South Korea in 1977 to revitalize its East Asian collections, while the National Arboretum sponsored a fresh round of thirteen plant-gathering ventures, mainly to Asia, by teams from multiple U.S. institutions between 1976 and 1989. These visits included the resumption of botanical exchanges with China in 1978 after thirty years' hiatus during civil war and state formation in the People's Republic.[95] Among the travelers to Japan between 1981 and 1986 was Roland M. Jefferson (b. 1923), the first African American botanist on staff at the National Arboretum. In the United States planting flowering cherry trees had fallen out of fashion for several decades, but Jefferson helped spur a fresh wave of popularity by collecting almost a half million seeds in Japan, which enabled genetic diversity in the stock of Japanese cherries in America. The Yokohama Nursery Company coordinated the gift of 3,000 cherries by Tokyo Mayor Ozaki Yukio to Washington, D.C., in 1912 and participated in the reciprocal gift of 3,000 dogwoods by the United States to Japan in 2012 to mark the centennial of the original donation. In 1980 Jefferson provided Watanabe Takeo, head of the Tokyo park system, with 2,000 cuttings representing eight of the twelve varieties of cherries donated in 1912, to replace cherries lost in Tokyo when the Ara river changed course.[96] In this way Jefferson was more responsible than any other single plant biologist for creating the beautiful current collections of flowering cherries in both Japan and the United States. He and other modern scientists ushered Japanese plants into the contemporary moment, long after the age of imperial botany in the mid-nineteenth century and the boom in Japanese gardens in Europe and North America early in the twentieth. Today plants from Japan are so common and so highly valued worldwide that they scarcely seem Japanese at all.

Four

Foundations of Plant Biology in Modern Japan

The Meiji Japan (1868–1912) from which Europeans and Americans gathered plants was an era of epistemological as well as political, industrial, and social transformation. Spearheaded by intellectuals such as educator Fukuzawa Yukichi (1835–1901), philosopher Nishi Amane (1829–1897), and legal historian Katō Hiroyuki (1836–1916), scholars, schoolteachers, and the new Ministry of Education rapidly set aside the Neo-Confucian theories of knowledge that prevailed in the Edo period in favor of European positivism, utilitarianism, and pragmatic approaches to learning based on inductive logic and laboratory investigation. Although materia medica (*honzōgaku,* herbal medicine) in the Edo era was more data-driven than theory-driven, and thus fairly compatible with modern science as practiced in Europe and North America, nonetheless its basic text, *Bencao gangmu/honzō kōmoku* (Outline of medical herbs, 1596), cast a stiff template within which most materia medica scholarship took place before the Meiji restoration of 1868. Japanese specializing in biology in the Meiji period, like their counterparts in the physical sciences, turned increasingly to Europe and North America for guideposts in conducting their research. For botany this meant a rapid transition from plant description and illustration to European principles of taxonomy and morphology, as taught both at European research institutes and at universities in the United States.

This chapter examines the development of research and instruction in the botanical sciences in Japan from the 1870s to the 1930s, with particular focus on the University of Tokyo (founded 1877, renamed Imperial University starting in 1886, Tokyo Imperial University in 1897, and University of Tokyo in 1947) and its renowned botanical garden at Koishikawa. It takes up transitional figures who shifted from *honzōgaku* to modern plant biology, such as the pioneer Itō Keisuke (1803–1901), then analyzes the roles of specialists in taxonomy and morphology, such as Yatabe Ryōkichi (1851–1899), Matsumura Jinzō (1856–1928), and Miyoshi Manabu (1862–1930), the latter three serving one after another as director of the Koishikawa Botanical Gardens. The chapter also considers research conducted at Sapporo Agricultural College (today Hokkaido University) and Kyoto University, as well as the roles of academic societies, scholarly journals, botanical gardens, arboretums, and herbaria in advancing knowledge of botanical science and Japan's distinctive plant life. Thanks to these and other forerunners, by the end of the Meiji period Japanese scholarship in botany earned international respect concurrent with the huge popularity of Japanese flora in gardens and landscapes in many parts of the world.

Botanical Beginnings

The *honzō* scholar Itō Keisuke (1803–1901) exemplified the transition from materia medica to modern plant science in Japan, as explained in chapter 2, and during the 1870s the Ministry of Education's Office of Natural History was the key vehicle for introducing European methods of studying plant life. Tanaka Yoshio (1838–1916), a student of Itō's at the Institute for Investigating Foreign Books (Bansho Shirabesho) in the 1860s, championed European natural history in his role as translator of Western botanical works for the ministry. He translated and adapted writings by the leading Swiss botanist Augustin Pyramus de Candolle (1778–1841) and his son Alphonse Pyramus de Candolle (1806–1892), supplemented by insights Tanaka drew from the work of Scottish botanist John Hutton Balfour (1808–1884). Unlike his mentor Itō, Tanaka favored applied uses of natural history and organized an exhibit of artifacts in Tokyo's Kanda district in 1872: "In my career," he later wrote, "I had the desire of showing things to people in exhibits and museums."[1] He followed out his passion for practicality by publishing a seven-volume work, *Yūyō shokubutsu zusetsu* (Illustrated useful plants, 1891), with abundant drawings and descriptions of Japanese flora.[2] Ono Motoyoshi (1838–1890), a grandson of the *honzō* scholar Ono Ranzan (1729–1810), like Tanaka studied medicine and materia medica, then joined the Ministry of Education's Office of Natural History and introduced the term *physiological botany* from English researcher John Lindley (1799–1865). Ono and Tanaka pioneered the use of *shokubutsugaku* for botany; the term has been standard since. Ono also published a Japanese edition of *Zhiwu mingshi tu kao* (Botanical terms with illustrations, 1848) by the Chinese official Wu Qijun (1789–1847) as *Shokubutsu meijitsu zukō* (Illustrated plant names, 1883–1884), the last significant botanical work produced in China before European botany was introduced to Chinese scholarship. Another transitional figure was Kaku Hika (1816–1894), who studied medicine and materia medica before moving to Tokyo in 1878 to work as an assistant to Itō at the Koishikawa Botanical Gardens, where he trained as a botanist and developed expertise on lilies and grasses.[3]

Formal instruction in Euro-American biology began in 1877 at the newly opened University of Tokyo when Edward Sylvester Morse (1838–1925) began a three-year professorship in zöology. Morse was a marine biologist and former lecturer at Harvard University who while in Japan became an amateur archaeologist and avid ceramist. He was soon joined at the University of Tokyo by Charles Otis Whitman (1842–1910), an ornithologist with a Ph.D. from the University of Leipzig who served as a professor of zoology between 1879 and 1881. At the University of Tokyo he offered upper-level courses and taught laboratory methods using a microscope. As one specialist put it, "Morse laid the rails, Whitman ran the train on them."[4] Morse's lectures evidently provided some material on botany, but his main contribution, like Whitman's, was introducing modern animal biology.[5] A prominent moment in the history of plant biology in Japan also occurred in 1877 with the arrival of a recent Cornell University alumnus, Yatabe Ryōkichi, as a professor of botany at the University of Tokyo, the first such appointment in the country's history. Almost immediately Yatabe was joined by the veteran scholar Itō Keisuke, who served for a decade as a visiting professor at the Koishikawa Botanical Gardens under Yatabe's direction, now that the garden was administered by the university. Itō did no teaching but shared his vast knowledge of plants with scholars in person and through the catalogs and research publications he edited.[6]

With Yatabe's appointment the University of Tokyo took the first steps

Figure 4.1 Former headquarters of the Tokyo Medical College (est. 1872), at the western end of Koishikawa Botanical Gardens, Tokyo, 2017.

Figure 4.2 Yatabe Ryōkichi (1851–1899). Photo by Ogawa Kazumasa. Public domain.

toward becoming Japan's premier institution for botanical studies, but not until 1895 did the university recognize botany as a discipline separate from natural history at the main campus and medical faculty. When the Imperial College (Kaisei Gakkō, est. 1871) and Medical College (Igakkō, est. 1872) merged to form the University of Tokyo in 1877, its Faculty of Science employed a dozen foreign and three Japanese scholars, with biology taught to upper-level students. Because Morse and Whitman were Americans and Yatabe was trained abroad, all botany and zoology lectures were initially delivered in English. This temporary reliance on foreign pedagogy was a major factor driving Yatabe's fierce determination in the 1890s to force botanical study in Japan to stand on its own footing, independent of international tutelage.

Yatabe learned English as a youth from Nakahama Manjirō (1827–1898), who was shipwrecked in 1841 and taken to the United States for ten years before returning to serve as a government translator and interpreter.[7] Yatabe became an instructor in botany at the forerunner of the Imperial College in 1869, then was sent to Cornell University with Japanese government financial support in 1872 for further study of plant science. After earning his B.Sc. in 1876 he did summer research at Harvard University under the mentorship of botanist William Gilson Farlow (1844–1919), then returned to his post at the Imperial College before being named a professor at the newly established University of Tokyo in 1877. There he trained a generation of young botanists, wrote new-style poetry (*shintaishi*), and became a prominent intellectual at the short-lived Rokumeikan (Deer Cry Pavilion, 1883–1888), a club where elite Japanese mingled with foreign diplomats and businesspeople. He reportedly "fell into idleness"[8] in 1890 and was forced to resign from the university the following year, then headed the Tokyo Higher Normal School (Tokyo Kōtō Shihan Gakkō) until he died in 1899 while swimming at Kamakura.[9]

Known as stubborn and unwilling to listen to others, Yatabe more than any other individual built the edifice of modern botany in Japan. He moved Japanese specialists beyond herbal medicine to the study of all plant forms. But the first years at the University of Tokyo were difficult. Sometimes only one or two students appeared for instruction in botany; an exceptional year saw five or six.[10] Yatabe's most important pupil was Matsumura Jinzō, who joined the faculty at the university in 1883 as an assistant professor after completing his training under Yatabe's direction. Another newcomer as assistant professor that same year was Ōkubo Samurō (1857–1924), who spent five years learning botany at the University of Michigan and two years of further study in Great Britain. Together with Itō Keisuke at the Koishikawa Botanical Gardens, these four scholars formed the nucleus of research in plant science at the university for the remainder of the 1880s.[11] Among the most promising young researchers under their direction were Miyabe Kingo (1860–1931), soon to become a professor at Sapporo Agricultural College, and Makino Tomitarō (1862–1957), the most prolific Japanese plant researcher in modern times. Although preoccupied with program building, mentoring students, and directing the botanical garden, Yatabe translated a textbook by Asa Gray into Japanese and published the three-part *Nihon shokubutsu zukai* (Japanese plants illustrated) in 1893. This profusely illustrated work was based on the collections at the Koishikawa Botanical Gardens. His *Nihon shokubutsuhen* (Japanese plants), incomplete at his death, appeared in 1900 but covered just 427 species.[12]

One of Yatabe's lasting contributions to plant biology in Japan was spearheading the Botanical Society of Tokyo, founded in 1882 and renamed the Botanical Society of Japan in 1931. Yatabe and Edward Morse established the University of Tokyo Biological Society in 1878, but botanists split off four years later to form the Tokyo Shokubutsu Gakkai; zoologists reconvened as the Tokyo Zoological Society (Tokyo Dōbutsu Gakkai) in 1885, later renaming it the Japan Zoological Society. Under the leadership of Ōkubo Samurō, in 1887 the Botanical Society of Tokyo began publishing an English-language journal of international repute, *Botanical Magazine, Tokyo* (*Shokubutsugaku zasshi*), retitled *Journal of Plant Research* in 1955.[13] Near the end of the twentieth century Japan counted a dozen major scholarly organizations dealing with plant research, and thirty-seven others with some engagement with botany.[14] At its founding in 1882 the Botanical Society of Japan proclaimed, "This society's purpose is to do everything in botanical research"—taxonomy, morphology, physiology, physical chemistry, cytology, genetics, ecology—not so different from today. By 1896 its journal was well received internationally, as one of its official publications wrote in 2016: "The Botanical Society was finally attaining world standards."[15] Today the focus has shifted from "whatever I research is interesting" to "how should I research something?" that is, process and method, not just content[16]—with a focus on formulating and testing hypotheses about plant biology.

"A Revolution in East Asian Botany"

Itō Tokutarō (1865–1941), a grandson of Itō Keisuke, spent the years 1884 through 1887 studying plant science in Cambridge, England, then returned the following year to Japan and called for "a big revolution in East Asian botany."[17] Whereas Itō remained a naturalist like his grandfather, it was Yatabe Ryōkichi who championed a real breakthrough in 1890 by calling on Japanese botanists to do their own plant identifications, no longer deferring to specialists abroad. "When I began to make collections of plants for the Imperial University of Japan [University of Tokyo], there was not a single reliable specimen in its museums, and, worse than this, no books of reference whatever." He turned to European and American botanists, "from some few of whom I received satisfactory answers to my inquiries; from others I received but scanty information; [from the rest] I have been so unfortunate as to have received no answer whatever even after the lapse of several years." But now that he had a collection of specimens and books, "I have decided to begin to give new names to those plants which I consider as new, without attempting in many cases to consult with European specialists. . . . The species to which they are given may not be wholly new, though new to the flora of Japan."[18] If not quite a declaration of independence, this bold statement of diminished reliance on foreign authority urged colleagues in Japan to boost their self-confidence and contribute their findings to the international community of scholars using nomenclature of their own choosing. Later in 1890 Yatabe frankly declared that "Japanese plants should be given scientific names by Japanese,"[19] a kind of botanical nationalism he obeyed by devising a new name for a variety of primrose, *Primula nipponica* sp. *nov.*, now known in his honor as *Primula nipponica yatabe.*

Matsumura Jinzō, Yatabe's most important pupil and colleague during the late 1870s and 1880s, followed an unconventional path from horticultural assistant to Japan's leading botanist by the turn of the twentieth century. Matsumura began working at the Koishikawa garden in 1877 with no previous training in plant science, but he quickly learned the principles of botany from Yatabe and by 1883 was appointed assistant professor at the University of Tokyo. Among his many duties was building the collections at the University of Tokyo herbarium, where he methodically identified new species and corrected many errors in previous classifications. Matsumura was an avid field biologist, walking the meadows and mountain slopes to identify and record new plant varieties. In 1884 he edited *Nihon shokubutsu meii* (Japanese plant names), adding binomial scientific names to familiar terms for Japanese plants and in the process setting aside the authority of Edo-period materia medica classifications in favor of taxonomy based on evolutionary principles.[20] For a time other scholars such as Ōkubo Samurō, Makino Tomitarō, and Miyabe Kingo embraced Matsumura's system, but subsequent generations of specialists changed many of the names used in his *Meii*, reducing the long-term impact of his usages and prompting a sympathetic biographer to comment, "We have to say this was truly a shame."[21]

Between 1885 and 1888 Matsumura studied taxonomy rather desultorily in Germany, first at Heidelberg, then at Würzburg, but without gaining much proficiency in the German language. Matsumura wrote his father from Würzburg on June 15, 1886, "I am surprised to see how Germany could produce so many great scholars, even though German students seem to be quite delinquent and lazy."[22] Returning to the Imperial University in 1888, he taught taxonomy and lectured widely on plant physiology, ecology, cytology, morphology, biochemistry, and genetics. Matsumura was promoted to full professor in 1890 and devoted the next decade to fortifying the foundations of Japanese botanical research in the wake of Yatabe's resignation from the university in 1891.[23]

Even more so than Yatabe Ryōkichi, Matsumura is regarded as the architect of modern plant science in Japan by designing both Tokyo Imperial University's botany laboratories and its herbarium, as well as training such top leaders of the next generation as Miyoshi Manabu, Koizumi Gen'ichi, and Fujii Kenjirō. His *Teikoku shokubutsu meikan* (Directory of plants in the empire, 2 vols.), published by Maruzen in 1904–1912, summarized Japanese botanical research to that point and provided synopses of floral surveys. Its Latin edition, *Index Plantarum Japonicarum*, appeared in two volumes from the same publisher in 1904–1905 and made Matsumura's scholarship known around the world. Partly as a result, he was honored by scholarly societies in Germany, Scotland, and Czechoslovakia and appointed a member of Japan's Imperial Academy; he lectured both the Meiji and Taishō emperors on botany. His interests reached beyond taxonomy and morphology to plant physiology, genetics, biochemistry, and ecology, prompting his best students to deepen these fields in their own work. Matsumura was also a skilled cataloger and is recognized today as namer of about twenty plant species, including the Somei Yoshino cherry (*Prunus yedoensis matsum.*). His *Nihon shokubutsu meii* (1884) is still a standard source for identifying specimens.[24]

With Matsumura's guidance Tokyo Imperial University took steps to integrate botanical studies of crops and forestry with the main university campus at Hongō. In 1890 the university merged with Tokyo School of Agriculture and Forestry (Tokyo Nōrin Gakkō) at Komaba to form the College of Agriculture, which in 1919 was renamed the Faculty of Agriculture and later moved to its present site at Hongō. Shirai Kōtarō (Mitsutarō, 1863–1932) was appointed to its first chair in agricultural botany in 1893. The College of Agriculture focused on physiology, microbiology, plant diseases, and diseases of aquatic animals. For its part, the College of Science counted 76 graduates in zoology and 56 in botany between 1881 and 1912. Many of these alumni became officials in various government agencies; a few completed graduate studies and fanned out to institutions throughout the country to teach and conduct biological research. The year 1912 marked the total separation of botany from the dominant zoology curriculum at the university, allowing botany to establish a more prominent disciplinary profile, but after World War Two zoology and botany came back together in the present Department of Biological Sciences.[25] In sum, the study of plant life in Meiji Japan, once marginal in and out of the academy, underwent considerable expansion and increasing self-identity during the first two decades of the twentieth century.

Another pupil of Yatabe's who rose to prominence in plant taxonomy and pathology was Miyabe Kingo, an early graduate of Sapporo Agricultural College (est. 1875) whose classmates included Christian leader Uchimura Kanzō (1861–1930) and educator Nitobe Inazō (1862–1933). Miyabe spent the years 1881 through 1883 as a researcher in the University of Tokyo's botanical laboratories learning how to classify Japan's flora, then was hired in 1883 as an assistant professor of botany at his alma mater in Sapporo, where he was tasked with establishing a botanical garden that today is one of Japan's finest. He served as a visiting researcher at Harvard University from 1886 to 1889 under the direction of Asa Gray and William Gilson Farlow, developing expertise in fungi and algae. Miyabe was promoted to full professor in 1889 and confirmed as the botanical garden's first director the following year.[26] The botanical garden opened in 1886 and began collecting plants from Hokkaido and offshore islands to the northeast; included were useful plants such as crops and ethnobotanical specimens from Ainu communities. From the outset the garden emphasized public service and encouraged visitors of all ages.[27] Starting in 1884 Miyabe became the first botanist to collect specimens along the Hidaka coast of Hokkaido, an area thick with first-growth forest dotted with occasional fishing spots, where he discovered a number of new species, including fungi and marine plants. His research was summarized in the twenty-eight-volume *Hokkaido shuyō jumoku zufu* (Illustrated main trees of Hokkaido), edited in collaboration with Kudō Yūshun (1887–1932) and completed in 1931.[28] Miyabe's work opened up the field of botanical studies in Hokkaido and guided subsequent scholars in identifying new or rare plants indigenous to the island. The Sapporo Agricultural College, which in 1907 was renamed the College of Agriculture of Tohoku Imperial University in Sendai, became Hokkaido Imperial University in 1918 and Hokkaido University in 1947. Despite Miyabe's pathbreaking research, program administration, and mentoring of junior scholars, botany became a separate department at the university only in 1930, after which three new professorships were added in short order.[29] Nonetheless Sapporo ranked with Kyoto Imperial University (est. 1897) and Tokyo Imperial University as one of Japan's three most important centers of botanical scholarship in the first half of the twentieth century.

Best-known of all the young scholars who gathered at Yatabe's and Matsumura's laboratories in the 1880s was Makino Tomitarō, an autodidact from a wealthy sake brewing family in rural Tosa (Kōchi prefecture) who collected 400,000 botanical specimens and identified nearly 1,600 species throughout his career. He almost single-handedly established the field of public botany in Japan and eventually received the degree of doctor of science from Tokyo Imperial University. After his death in 1957 the government recognized his achievements by posthumously conferring the order of culture. His dried plant specimens were donated to the new Makino Herbarium at Tokyo Metropolitan University (Shuto Daigaku Tokyo) in 1958; the Makino Memorial Garden and Museum were opened that same year in Higashi Ōizumi, Tokyo; and the Kōchi Prefectural Makino Botanical Garden was also established in Kōchi city in 1958.[30] His attainments were featured in two exhibits at the National Museum of Nature and Science in Tokyo's Ueno park in 1998 and 2012–2013. He is known worldwide because "Makino" appears in the scientific names of dozens of the many plant species and varieties he discovered throughout his long career.

Makino quit school at the age of fifteen and began teaching himself botany by surveying and cataloging plants in western Kōchi prefecture. In 1884 he traveled by ship to Kobe and by steam train onward to Tokyo, soon gaining an informal affiliation with scholars at the University of Tokyo: "At the botany laboratories were three professors, Matsumura Jinzō, Yatabe Ryōkichi, and Ōkubo Samurō. Because a man [Makino] had arrived who was enthusiastic about the excellent plants in the deep mountains of Shikoku, these professors welcomed me warmly."[31] The university's science units at this time were housed in a white, two-story wooden building at Hitotsubashi in Tokyo, as were the modest botanical laboratories. Together with Matsumura and Ōkubo he was soon invited for dinner at Yatabe's home, despite his lack of formal schooling.[32] But within two years Yatabe showed displeasure with Makino, upbraiding him for not publishing the results of his research; Makino, in turn, criticized some mistranslations of terms in Yatabe's Japanese-language edition of a textbook by Asa Gray. After this contretemps Yatabe discouraged Makino from using the botanical laboratories, yet the following year he published Makino's article on Japanese pondweed in the inaugural number of *The Botanical Magazine, Tokyo* (1887).[33]

Yatabe was further displeased when Makino, at his own expense, published the first volume of *Nihon shokubutsu shizuhen* (Illustrated Japanese plants, 1888) rather than having it appear from Tokyo Imperial University, where he did the bulk of the research. This volume consisted mainly of illustrations by Makino himself, drawn at a level of detail almost "photo-like in accuracy."[34] He set aside Chinese plant names in favor of those in Japanese kana, a practice later adopted by Matsumura Jinzō, who sidestepped Yatabe in endorsing Makino's book: "Makino Tomitarō is the only person in the Japanese empire who could write an illustrated volume on the plants of our country."[35] After Makino presented a report in 1890 on his discovery of an insect-eating waterwheel plant (*Aldrovanda vesiculosa*) on the banks of Tokyo's Edo river, Yatabe—perhaps threatened by Makino's evident talent as well as his independent behavior—told him, "You and I are thinking of publishing separate [competing] books on Japanese plants, so from now on you're forbidden to see the books or specimens in the laboratory" at the university.[36] Cashiered henceforth from the College of Science, Makino found a nominal spot in the research laboratory of the university's College of Agriculture at Komaba (today the Faculty of Agriculture at the Hongō campus), but he lacked the resources to publish further volumes of *Nihon shokubutsu shizuhen.* Stymied by his differences with Yatabe, whose elite background was at odds with Makino's small-town origins, Makino carried on with his plant collecting, botanical surveys, interaction with naturalists all over Japan, and scholarly writings as a virtual lone scholar for two years beyond Yatabe's forced departure from the university because of an internal power struggle in 1891.[37]

In some ways the self-educated Makino exemplified Japan's era of "rising in the world" through individual ambition (*risshin shusse*) and self-betterment, yet he certainly benefited from his connections, however tenuous at times, with the nation's most prestigious university. Through access to the specimens gathered by Yatabe and Matsumura and the research methods they mastered while abroad, Makino's career advanced far more rapidly than if he had been an entirely independent scholar in the 1880s and early 1890s. His exile from the university ended in 1893 when Matsumura, the successor to Yatabe as laboratory head and botanical garden director, appointed Makino to lecture on botany as assistant (*joshu*), a regular appointment with a salary of fifteen yen per month, equivalent to $7.50 in 1893 U.S. dollars (Matsumura's salary was fifty yen per month).[38] To supplement his income, in 1893 Makino accepted the invitation of Hamao Arata (1849–1925), president of the university and a leading politician, to edit *Dai Nihon shokubutsushi* (Greater Japan botanical journal), published by the Hongō campus. Matsumura, always correct but seldom warm in his dealings with Makino, apparently disagreed with Makino's leadership of the journal, which ceased publication after just four volumes. Moreover, Makino assigned names to plants that often contradicted those in Matsumura's 1884 *Nihon shokubutsu meii*, prompting some level of resentment. Makino later wrote that "Professor Matsumura exerted oppression on me both personally and in scholarly terms," yet Makino kept on researching and publishing, which "invited Professor Matsumura's displeasure with me."[39] Matsumura's biographer Nagakubo Hen'in considers Matsumura's criticism of Makino "fleeting"[40] and dismisses Makino's repeated claims of being bullied by Matsumura as "a defect in Makino's character."[41] The spat led to continuing coolness between the two until Matsumura's death in 1928.

Japanese scholars in the early decades of modern plant biology sought to project a distinctive image, led by Yatabe's insistence in 1890 on naming their own discoveries, Makino's example of a career launched without study abroad, and pride in the nearly 6,000 species of higher plants that gave the Japanese islands such enviable botanical diversity. Yet they also reached out for acceptance and validation from the world community of scholars, most notably seeking international legitimacy by publishing their research in English-language journals of wide circulation, starting with *The Botanical Magazine, Tokyo* in 1887. A signal moment came in 1896 when a pair of research affiliates at Tokyo Imperial University independently discovered the presence of swimming sperm cells in two seed-bearing trees, ginkgos and sago cycads. Hirase Sakugorō (1856–1925), an assistant at the university's botanical laboratories in the Koishikawa garden, announced in the October issue of the magazine his findings about a ginkgo long planted in the garden. Swimming sperms were well known in ferns and mosses, but Hirase's was the world's first discovery of this phenomenon in a seed plant, overthrowing the scholarly consensus among morphologists that plants propagated by seeds had no sperms. His research built on a good deal of concurrent study of ginkgos at Koishikawa; Fujii Kenjirō (1866–1952), a geneticist appointed in 1895 as an assistant in the university's College of Science, published work on the morphology of ginkgo flowers in

1896 and expressed the growing consensus that ginkgos might have evolved from fernlike ancestors.[42] One month after Hirase's discovery reached print, Ikeno Seiichirō (1866–1943), at assistant professor at the university's College of Agriculture at Komaba who was married to Matsumura's younger sister, published his findings about sperms in sago cycads that he had studied in Kagoshima prefecture. Ikeno's conclusions gave further credence to the view that ginkgos and sago cycads, often erroneously called fern palms, differ in appearance but actually resemble each other botanically.[43]

Both Hirase and Ikeno conducted their research in Japan without scholarly cooperation from Europe or America, a source of astonishment abroad and considerable self-congratulation among scholars at home. Miyoshi Manabu, appointed as a full professor of botany at Tokyo Imperial University in 1895 after returning from study in Leipzig with plant physiologist Wilhelm F. P. Pfeffer (1845–1920), expressed hopes of achieving the revolution in Japanese plant sciences advocated by Itō Tokutarō in 1888. Miyoshi was excited by Hirase's and Ikeno's discoveries and offered high praise in a review published in the December 1896 issue of *The Botanical Magazine, Tokyo:* he called Hirase's findings "exceedingly interesting and important work, in no sense miraculous" but instead the product of careful research. Miyoshi credited both scholars with "achieving important results for botanical taxonomy and also for plant physiology."[44] The broad international impact of the 1896 discoveries was evident when the British paleobotanist and women's rights advocate Marie Stopes (1880–1958), drawn in part by Ikeno's work on the reproduction of sago cycads, spent eighteen months at Tokyo Imperial University in 1907–1908. The Imperial Academy (Teikoku Gakushiin) honored Ikeno, and, at Ikeno's insistence, Hirase as well, with imperial prizes in 1912.[45] Later generations of Japanese scholars continued to regard the Hirase and Ikeno discoveries as "epoch-making" (1973)[46] and as events that "created a stir in worldwide botany" (2016).[47] The steady flow of international researchers arriving in Japan beginning in the early twentieth century indicated their increasing regard for Japanese plant biologists as well as their fascination with Japan's huge range of vegetation. The influx included K. K. Tsoong (1868–1940), who became one of China's leading botanists in the Republican period (1911–1949), and historians Wang Rongbao (1878–1933) and Ye Lan (1875–?), both of whom studied botany at Waseda University in 1901.[48] These researchers were among the hundreds of thousands of Chinese, Koreans, Taiwanese, and Southeast Asians who flocked to Japan during the first two decades of the twentieth century to study science, medicine, engineering, and technology as well as the humanities and social sciences, in search of a new Asian modernity that was not specifically Euro-American in provenance.[49]

Infrastructure for Research

As with most areas of modern scholarship, Japan began the Meiji era with inadequate physical resources for botanical research—libraries, laboratories, museums, arboretums, botanical gardens, herbaria—but with a rich endowment of books and illustrations generated by specialists in materia medica during the Edo period, as well as a sizable number of herb gardens for medical plants established by Japan's early modern domains. Another asset was an inheritance of scholarly curiosity and rigorous schooling in private academies throughout much of the country. The nation's eagerness to gain new knowledge after the encounter with Western gunboat diplomacy in the 1850s outran the infrastructure to support this thirst for learning in the biological sciences no less than in most other fields of scholarship, a condition that persisted for several decades after the Meiji restoration in 1868.

For many years after its founding in 1877, the National Science Museum (now called the National Museum of Nature and Science), located in Tokyo's Ueno park since 1931, was the chief institution charged with collecting, displaying, and interpreting artifacts concerning natural history. Thoroughly renovated in recent years, the museum presents permanent displays on Japan's prehistory as well as special exhibitions on science and technology. An Institute for Nature Study, established in 1949 in Shirokanedai, Tokyo, for the study of ecology on the site of a former imperial forest, was added to the museum in 1962, and the Tsukuba Botanical Garden in Ibaraki prefecture opened in 1983 as the science museum's main research facility for the study of plants. Today Japan counts more than 400 natural history museums, most of them small; only 14 of them were built before 1945.[50]

Botanical gardens and arboretums were few in number throughout the Meiji (1868–1912) and Taishō (1912–1926) periods. In addition to the venerable garden at Koishikawa, which was renamed Koishikawa Botanical Gardens in 1875, Shinjuku botanical garden was established in 1872 on the grounds of the former Naitō daimyo estate and renamed an imperial garden in 1879, but with restricted use for research. This fifty-eight-hectare property was opened to the public as Shinjuku Gyoen in 1949 and today contains 1,500 trees representing seventy-five species of cherries and eight hundred varieties of chrysanthemums.[51]

The Sapporo Botanical Garden (est. 1886), Osaka's Tennōji park (1909), the Tama Forest Science Garden (Tama Shinrin Kagakuen, 1921) in suburban Hachiōji, and the Kyoto Prefectural Botanical Garden (1924) were among the few major research facilities for botanical study before World War Two. Many so-called botanical gardens were built after 1945, but most are little more than public parks. As plant biologist Iwatsuki Kunio pointed out in 2004, "Even

Figure 4.3 Shinjuku Imperial Garden became a public park in 1949 after a half century as an imperial recreation retreat in the heart of Tokyo, 2009. Skyscrapers began to loom over the park in 1971 on the site of the former Yodobashi reservoir west of Shinjuku station.

today in attaining botanical gardens as facilities for botanical study [Japan is] very far behind Western Europe."[52]

Koishikawa Botanical Gardens began in 1638 as a medicinal plant nursery for the Edo government and underwent many changes before becoming a unit of the new Ministry of Education after the Meiji restoration. Within a few years it became the responsibility of the city of Tokyo, then in 1875 it affiliated with the predecessor to the National Museum of Science and assumed its status as a botanical garden, no longer intended for materia medica. Its new function as a neighborhood park as well as locus for studying plants was reflected in a sudden jump in attendance from 738 visitors in 1875 to 3,237 the following year.[53] In 1877 it was transferred to the newly founded University of Tokyo's College of Science as its chief botanical research facility, with Yatabe Ryōkichi as director and Itō Keisuke as adjunct professor at the botanical garden. Itō's botanical knowledge, especially of medicinal plants, was formidable, but because he lacked a university degree he was appointed as an *ingai* (nonmember) professor, equivalent to adjunct, the only professor ever to hold that status at the university, a situation rectified in 1881 when he was promoted to full professor, with research duties only.[54] The botanical garden underwent many changes of name within the university's structure across the

Figure 4.4 Tama Forest Science Garden on the slopes of Mount Takao in Hachiōji, Tokyo, contains an extensive collection of cherry trees and a towering stand of metasequoias, 2017.

next century, and even today it is technically considered a facility attached to the Graduate Faculty of Science at the University of Tokyo. One of the garden's distinctions was publishing in October 1877 volume one of *Koishikawa Shokubutsuen sōmoku mokuroku* (Catalog of plants in the Koishikawa Botanical Gardens, 2 vols., 1877–1881), "said to be the University of Tokyo's first scholarly publication."[55] Another distinction is that its collections included plants newly gathered in both Japan and Korea beginning with 1876, the same year the Meiji government imposed the Treaty of Kanghwa (Ganghwa) on the royal Korean government, the first in a series of steps to incorporate Korea into the Japanese empire, culminating in annexation of the entire peninsula as a colony in 1910.

When Matsumura Jinzō succeeded Yatabe as director of the Koishikawa garden in 1893 he began to move the botanical research laboratories from

Figure 4.5 Koishikawa Botanical Gardens, Tokyo, 2017.

their cramped quarters at the main university campus in Hongō to the botanical garden, making Koishikawa the true center of advanced training in plant biology in Japan. Instructional laboratories in botany split off from zoology and shifted from Hongō to a wooden building at Koishikawa four years later. An important new facility was opened in 1902 at Nikkō, later greatly expanded in 1911, for studying Japan's wide range of alpine plants.[56] Like many parks and open areas throughout the capital region, the Koishikawa Botanical Gardens suffered great damage in the wake of the Kanto earthquake and fires on September 1, 1923, especially because 35,000 evacuees from the devastation fled there, cutting down trees for shelter and firewood. Ōba Hideaki, Japan's leading botanical historian, has written that nearly all the bamboo at Koishikawa was felled and sharpened into spears by the refugees because of unfounded rumors that Korean residents of Tokyo were causing disturbances.[57] The next month twenty-two barracks were erected to house evacuees still living in the garden; by May 1924 all had departed, and the barracks were torn down. In the summer of 1934 the botanical laboratories were moved back to Hongō with the completion of a second building for the Faculty of Science at Tokyo Imperial University.[58] Since then the botanical garden has continued to collect live specimens from East and Southeast Asia as well as other world areas. Today it contains about 4,000 species of woody, herbaceous, and tropical plants and a library of more than 20,000 volumes in plant biology. Koishikawa set a high scholarly standard for other botanical gardens in Japan, most of them established after World War Two, as discussed in chapter 6.

The University of Tokyo's herbarium collection of dried plants originated with the first instruction in botany in 1877, but it was slow to develop despite the importance of herbaria not only for taxonomy but also for botany as a whole. Like their Edo-period predecessors, many modern Japanese scholars—above all Makino Tomitarō—collected, cataloged, and housed their own private collections, rendering them all but inaccessible to the scholarly community at large. Tokyo Imperial University moved its herbarium specimens to the Koishikawa Botanical Gardens in 1893 when Matsumura Jinzō began shifting the botanical laboratories there; they stayed at Koishikawa until 1935, when they were moved back to Hongō. Today the collections are split between the main university campus and the botanical garden, totaling about 1.7 million specimens, with particular strengths in Asian and Himalayan plants. Maintaining fragile specimens is labor-intensive, housing rarely consulted examples of dried plants is costly, and Japanese scholars sometimes feel shortchanged by the relative paucity of herbarium facilities in contrast with the thriving state of botanical research in their country.[59] Still, in addition to collections at the University of Tokyo, National Museum of Nature and Science, and Makino Herbarium at Tokyo Metropolitan University, important herbaria are located at Hokkaido, Tohoku, Kanazawa, Kyoto, Kagoshima, and Ryukyu Universities as well as at the Osaka Museum of Natural History.

A significant addition to research infrastructure occurred in 1917 when the Institute of Physical and Chemical Research (Rigaku Kenkyūjo) was established by industrialist Shibusawa Eiichi (1840–1931). Privately funded and initially based at Komagome, Tokyo, the institute abolished its physics and chemistry divisions in 1922, became widely known by its acronym Riken, and set up a series of independent laboratories. In 1958 it was reorganized as a public corporation, and in 1967 it moved to larger quarters in Wakō, a suburb of Tokyo. Although some degree of botanical research occurred at the Riken of the 1920s and 1930s, especially regarding medicinal plants, the life sciences received a boost when Riken opened a satellite facility at Tsukuba, Ibaraki prefecture, in 1984. Two new centers with biology laboratories opened in Kobe (2000) and Yokohama (2003). In 2015 Riken was designated a National Research and Development Institute, with positive implications in the long run for improved financial support from the government after years of declining budgets (see chapter 6).[60]

Diversification and Specialization

Japan ended the nineteenth century with two important centers of botanical research, Tokyo Imperial University and Sapporo Agricultural College, both closely linked at first through Tokyo alumnus Miyabe Kingo, founder and first director of the college's botanical garden. By 1982 roughly one hundred universities and research institutions sponsored scholarship in plant biology, led by the University of Tokyo, Hokkaido University, and Kyoto University. The latter, founded in 1897 as Kyoto Imperial University, established a biology department in 1919 and appointed Koizumi Gen'ichi (1883–1953) as assistant professor of botany. Koizumi studied with Matsumura Jinzō at Tokyo Imperial University, earning his D.Sc. degree in 1913. While a student he lodged at the house of Matsumura's sometime rival, Makino Tomitarō, in Tosakichō, Koishikawaku. Koizumi later gained fame for his studies of plant life in Okinawa and led a research group that in 1932 began publishing a landmark serial, *Shokubutsu bunrui chiri* (Journal of phytogeography and taxonomy).[61]

The diversity of botanical research in Japan eventually spread far beyond established fields such as taxonomy and morphology, each of which also underwent considerable development. Much new taxonomy took place between 1900 and 1920, as well as new plant collecting in the Ogasawara (Bonin) islands, the Ryukyus, Taiwan, and less traveled regions of Japan. In plant physiology, Miyoshi Manabu trained a generation of researchers at Tokyo Imperial University and was author or coauthor of more than sixty studies, culminating in his *Sakura* (Cherries, 1938), a 467-page illustrated classic from the Tokyo publisher Fūzanbō.[62] Also at Tokyo Imperial University was Shibata Keita (1877–1949), who studied plant pigmentation and sponsored a great deal of research in metabolic physiology and biochemistry. Botanical historian Ōba Hideaki says of Shibata that "as Japan's pioneer in physiological chemistry, he laid the foundations" for this subfield of plant science.[63] Japanese advances in cytology occurred slightly later. Kamiya Noburō (1913–1999), who studied at Tokyo Imperial University, in Germany, and in the United States, researched the movement of protoplasm in cells of pure slime molds and stoneworts. In the postwar period Kuroiwa Tsuneyoshi (b. 1941) used a high fluorescence microscope to study molecular DNA and explain maternal heredity. He is best known for discovering the machinery of mitochondrial and chloroplast division.[64]

Fujii Kenjirō developed plant genetics at Tokyo Imperial University after returning from study in Germany and Great Britain in 1905. Johann Gregor Mendel (1822–1884) had published his findings about genetic inheritance in 1865 after eights years' research on pea plants; Mendel's laws were rediscovered in 1900 and soon led to research by Fujii in applied genetics, with a view toward breeding plants to increase agricultural output. In 1918 he was appointed to Japan's first chair in genetics through a gift to the university from the Osaka businessperson Nomura Tokushichi (1878–1945), founder of the Nomura financial group.[65] Kihara Hitoshi (1893–1986) was trained at Hokkaido Imperial University and in Sweden, then became a professor at Kyoto Imperial University and chair of the Ski Association of Japan. Kihara studied the genome of hybrid wheat and after World War Two helped to establish the National Institute of Genetics (Kokuritsu Idengaku Kenkyūjo). He also led the shift of Japanese plant biology to focusing on molecular genetics.[66] Japanese research in plant physiology and embryology was stimulated by the 1934 discovery of auxins, a class of hormones that regulate plant growth and differentiation. Sugiura Masahiro (b. 1936) graduated from Nagoya University, did advanced research at the University of Illinois and the University of California, San Diego, and as a professor at Nagoya identified all the genes in tobacco chloroplast.[67] Ōyama Kanji (b. 1939) and his research team did the same for Marchantiophyta (liverwort *Hepaticae*). Japanese scientists played a key role in the international genome project, identifying an entire rice plant genome in 2004. More recent research is ongoing into vegetable proteins.[68]

Bryology (study of mosses and liverworts) in Japan traces to the late nineteenth century at Tokyo Imperial University, where Yatabe, Matsumura, Miyabe, Ōkubo, and Makino all collected specimens for analysis. So well regarded was this subfield of botany that Shirai Kōtarō contributed an article, "Record of Experiments in Growing Bryophytes," to the inaugural issue of *The Botanical Journal, Tokyo* in 1887.[69] Twelve years later the journal announced the name of a new genus, *Makinoa*, named for Makino Tomitarō. At Tokyo Number Two Higher School biology teacher Yasuda Atsushi (1868–1924) and his students collected mosses, summarizing their findings in *Shokubutsugaku kakuron, inkabu* (Botany detailed, cryptogams, 1911), considered a key work for the taxonomy of Japanese mosses.[70] Iishiba Eikichi (1873–1936) summarized Japanese bryophyte research in two works that set a standard for the next generation, *Nihonsan tairui sōsetsu* (Introduction to Japanese liverworts, 1920) and *Nihonsan senrui sōsetsu* (Introduction to Japanese mosses, 1929).[71] Miyoshi Manabu was a pathbreaking scholar of lichens in Japan, contributing more than twenty articles on the subject to *The Botanical Magazine, Tokyo*. In 1893 he was the first Japanese scholar to describe a lichen with his identification of *iwatate*, long used in cooking, as *Gyrophora esculenta* (now usually *Umbilicaria esculenta*). Yasuda Atsushi succeeded Miyoshi as Japan's leading expert on lichens; he died suddenly in 1924, so his colleagues published his unfinished work the following year as *Nihonsan chiirui zusetsu* (Illustrated Japanese lichens).[72] Research on fungi began at the same time as that on mosses and lichens, initially conducted by Shirai Mitsutarō, Miyabe Kingo, and their pupils. One of the most innovating was Tanaka Nobujirō (1864–1905), who in 1890 announced his discovery of two new mushroom species, *hatsutake* (*Lactarius hatsudake*) and *akahatsutake* (*Lactarius akahatsu*). Tanaka later became expert in plant pathology, describing illnesses in purple-leaf mulberries.[73]

Although Miyabe Kingo had studied seaweeds briefly in the mid-1880s, the field of marine botany, especially algae and seagrasses, became a major area of research from the moment Okamura Kintarō (1867–1935) embarked on Japanese seaweed studies (algology) in 1889 after undergraduate training from Yatabe Ryōkichi, who had briefly studied seaweeds at Woods Hole, Massachusetts, in 1876. While a student Okamura lived in Fukagawa, Tokyo, near a seaweed (nori) farm, observing the growth of these plants and devoting himself to marine botany as a career. Okamura was a prolific scholar of oceanic algae who headed the Imperial Fisheries Institute (Suisan Kōshūjo, today Tokyo University of Marine Science and Technology or Tokyo Kaiyō Daigaku) from 1924 to 1931. In "Kaisō mandan" (Idle talk about algae, 1929) he wrote, "I was admitted to graduate school in 1889 . . . Yatabe sensei lent me specimens that were at the university," where there were only a half dozen books on algae. "Thinking back today, I'm surprised at how recklessly I began my work under such [spartan] conditions."[74] Okamura's most influential book was *Nihon kaisōshi* (Japanese marine plants, 1936), which described and illustrated 1,006 plants from Japan's shoreline, including a few freshwater species from riverine estuaries. Before Okamura, most knowledge of Japanese marine vegetation was attributable to foreign researchers; thanks to his four decades of investigations, far more became known about the development of Japan's marine plants than anyone had previously reported.[75] A dozen years after Okamura inaugurated Japanese studies of seaweeds, Endō Kichisaburō (1874–1921) graduated from Tokyo Imperial University's botany department and began a career studying sea mosses and the genus *Sargassum* (Sargasso grass) while serving as a professor at Hokkaido Imperial University. Okamura and Endō dominated academic study of marine plants for two decades, but many other specialists entered marine-products industries and contributed to botany through their work on commercial plant propagation.[76]

Plant ecology is often associated with the independent scholar and naturalist Minakata Kumagusu (1867–1941) in Wakayama prefecture, but Miyoshi Manabu first introduced this field to the academy when he returned to Tokyo Imperial University from Germany in 1895. At that point Miyoshi taught only general ecology or plant sociology. German zoologist Ernst Haeckel (1834–1919) first named the outlook of Prussian geographer and explorer Alexander von Humboldt (1769–1859) "ecology" in 1866, but the idea was slow to catch hold. Miyoshi had encountered ecology while studying in Germany from 1891 to 1895 and used the term *seitaigaku* (ecology) in his classroom lectures as early as 1895. He introduced the term *seitaigaku* to Japanese scholarship in 1903, shortly after publishing his most important work of experimental botany, *Shokubutsu seitai bikan* (Beauty of plant life, 1902).[77] Miyoshi also wrote *Shokubutsu no shakai* (Community of plants, 1903) and *Futsū shokubutsu seitaigaku* (Ecology of ordinary plants, 1908). His pupil Nakano Harufusa (1883–1973) was appointed an assistant professor of botany at the university in 1924 after researching tropical plants in Java and India, developing the subfield of tropical plant ecology and writing many studies of plant physiology and ecology. Kōriba Kan (1882–1957), who graduated from Tokyo Imperial University and became a professor of botany at Kyoto Imperial University, traveled widely from Mongolia and Siberia to the rain forests near Singapore to carry out research in plant physiology and ecology. As with other fields of biology, plant ecology broadened its scope greatly in the 1960s and became a mainstream field of scholarship in Japan.[78]

By the end of the Meiji period Japanese botanical studies had expanded from the introductory efforts of Yatabe Ryōkichi, Matsumura Jinzō, and Miyabe Kingo to a full palette of research specialties, many of them earning international respect thanks to publications in English and occasionally German. At the same time plant biology in Japan took an increasingly distinctive turn, foreshadowed by Itō Tokutarō's 1888 call for a discrete East Asian botany, Yatabe Ryōkichi's insistence two years later that Japanese scholars name their own discoveries, and especially the breakthrough research on ginkgos and sago cycads reported by Hirase Sakugorō and Ikeno Seiichirō in 1896. This distinctiveness became more apparent during Japan's age of empire in the early twentieth century, the topic of chapter 5.

Five

Plant Biology in Japan's Age of Empire

Historians commonly refer to the five decades between the victorious first Sino-Japanese war (1894–1895) and disastrous defeat in World War Two (1937–1945) as Japan's age of empire. The period from 1905 to 1931 was characterized by imperial democracy, a time of colonialism abroad and growing political participation at home by workers, women, farmers, and other previously excluded groups.[1] The years between 1931 and 1945 are best described as a time of imperial militarism, as Japan's forces seized Manchuria (northeast China), other parts of eastern and southern China, and major areas of Southeast Asia in an effort to construct a zone of economic self-sufficiency called the Greater East Asia Co-Prosperity Sphere. This chapter begins by considering the development of botanical studies in Japan during the first half of the twentieth century, then takes up the seldom-discussed topic of colonial botany, more aptly known as the botany of empire.

Makino and Minakata on Studying Nature

Looking back at the age of eighty-six over a career that coincided with the rise of modern plant biology as both an academic discipline and an amateur avocation, Makino Tomitarō wrote, "In a person's life, there is no benefit to being fond of nature. This is because people are a part of nature. Being friendly with nature first arises from feeling glad to be alive. To be fond of nature, first one has to renounce the self and plunge into nature . . . nature is our priceless teacher."[2] Although poles apart in scholarly impact, Makino and Minakata Kumagusu were strikingly similar in background, unconventionality, and outlook on nature: were he still alive in 1948, Minakata might well have endorsed Makino's view quoted above. Minakata was five years Makino's junior but belonged to the same scholarly generation; both hailed from prefectures far from the metropole (Makino from Kōchi, Minakata from southern Wakayama); neither was a college graduate. Both Makino and Minakata bucked the establishment; both were true naturalists with a genuine and lifelong love of plants; each certainly had an enormous impact on amateur botanists as well as professional scientists. Even though Minakata spent a fourth of his adult years overseas, Makino is far better known, both within Japan and abroad, chiefly because of his relentless outpouring of scholarly writings but also because of his leadership in public botany.

Among these publications was Makino's incomplete series *Dai Nihon shokubutsushi* (*Icones Florae Japonicae*, 4 vols., 1900–1911) with its precise text and skilled illustrations of a range of Japanese plants, "a real benchmark."[3] He also launched *Journal of Japanese Botany* (*Shokubutsu kenkyū zasshi*) in 1916, a key scholarly resource in phytotaxonomy as well as a beacon of public botany that is still published today. In the wake of Japan's victory in the Russo-Japanese war the Ministry of Education prioritized learning about nature in elementary schools, prompting a spate of textbooks in the next five or six years. In 1906 Makino collaborated with Murakoshi Michio (1872–1948) on a monthly illustrated publication for primary schoolchildren about plants; they also jointly produced volumes in 1908 on wild plants and plant illustrations, but thereafter they became competitors to finish major books. Makino worked steadily and fastidiously, whereas Murakoshi rushed to match his rival. As a result, Makino's *Nihon shokubutsu zukan* (Illustrated Japanese plants) and Murakoshi's *Dai shokubutsu zukan* (Great illustrated plants) appeared within a day of each other in 1925. The latter contained a preface from Makino's nominal boss at Tokyo Imperial University, Matsumura Jinzō, spurring a degree of resentment on Makino's part. Yet the rivalry was a charade because the audiences for the two books were different: Makino's was intended for specialists, Murakoshi's for general readers.[4]

Makino had already scaled back his position at Tokyo Imperial University to part-time in 1912 and thereafter gathered more herbarium samples for his private collection, no longer for the university herbarium. He named many new species, most of them not included among examples in the university's collection, and grew increasingly anxious about protecting his specimens of dried plants. A wealthy Kobe businessperson and art collector with a passion for plants, Ikenaga Hajime (1891–1955), read a story in the *Asahi* newspaper in 1916 about Makino's financial straits and decided to donate between ¥20,000 and ¥30,000 (U.S. $10,000 to $15,000 at the time) so Makino would not have to sell his specimens abroad. Ikenaga established the Ikenaga Botanical Research Institute in 1916 with an herbarium in Kobe intended to house specimens he bought from Makino, many of which had been collected by amateurs around the country. A majority were in poor condition, wrapped in newsprint without proper naming or indications of where and when they were gathered. Ikenaga's aid to the botanist proved to be sporadic and soon gave out.[5]

Although he never attended high school, college, or graduate school, Makino became so distinguished a scholar that Tokyo Imperial University conferred the degree of doctor of science in 1927, despite his misgivings about the honor stemming from touchy relationships years earlier with Yatabe Ryōkichi and Matsumura Jinzō. In 1932 Makino published *Kokuyaku honzō kōmoku* (Materia medica in Japanese translation), using headnotes at the top of the page to correct numerous mistaken Chinese plant names in use since Ono Ranzan's classic 1803 book on Chinese herbal medicine. Makino's seven-volume *Shokubutsugaku zenshū* (Complete botanical collection, 1934–1936) won a prize in 1937 from *Asahi,* which called Makino "the father of Japanese plants."[6] Nonetheless, amid protracted war in China, national mobilization at home, and a brewing crisis in financing higher education, the university forced Makino to retire in 1939 after forty-seven years' association with its research in plant biology.[7]

Makino's greatest achievement was *Makino Nihon shokubutsu zukan* (Makino's illustrated Japanese plants, 1940), a 1,213-page compendium that quickly sold out its first printing of 5,000 copies. Makino regarded his 1925 *Nihon shokubutsu zukan* as a "temporary book," brought into print prematurely because of competition with Murakoshi, that served as a sketch for the greatly expanded 1940 book. In compiling *Makino Nihon shokubutsu zukan* he was assisted by Nakai Takenoshin (1882–1952), a professor in the Tokyo Imperial University botany laboratories and director of the Koishikawa Botanical Gardens. The 1940 work was reprinted several times, expanded in 1956 by Maekawa Fumio (1908–1984) and others, and turned into an easy-to-read edition in 1961. A full-color version in three volumes appeared in 1982–1983, a handsome improvement over the black-and-white edition necessitated by wartime stringencies in 1940. Further additions and revisions of *Makino Nihon shokubutsu zukan* took place in 1989; today a student edition and a compact edition are for sale as well. As of 2014, the publisher Hokuryūkan estimated that the total number of copies sold in all editions was 400,000.[8] Despite the éclat that greeted his book, Makino faced chronic financial difficulties and relied on assistance from Fukuda Ryōtarō (1884–1956), president of Hokuryūkan. The Japanese government sent Makino to Manchuria in 1941 to conduct a census of cherry trees, and near the end of the war he chaired a panel, "Warning about Musashino Grasses," to help people in the Tokyo region identify inedible wild plants as they foraged for food amid severe commodity shortages and mounting malnutrition. He lectured the Shōwa emperor on plants in 1948 and kept adding newly discovered species to updated editions of his 1940 *Zukan.* At an advanced age Makino was elected to the Japan Academy (Nihon Gakushiin, formerly Imperial Academy), was awarded prizes for cultural merit, and was named an outstanding citizen of Tokyo before his death in 1957 at the age of ninety-five.[9]

Among Makino's many accomplishments was promoting the important subfield of public (amateur) botany beginning early in the twentieth century. Plant collecting, like bird watching, was a hobby for Japanese with the time and interest for rambles in fields and forests throughout the Edo period. One example is the self-taught naturalist Matsumori Taneyasu (1825–1892), a samurai from Tsuruoka in today's Yamagata prefecture who worked for the domain government and took up botany later in life. Matsumori produced an estimated 400 volumes of highly personal interpretations of natural history, including his idiosyncratic version of evolutionary theory. He had a keen eye for plants and provided excellent illustrations to accompany his scholarship, which is so well regarded that his major works are still in print today.[10] At the start of the twentieth century fourteen young naturalists at Tokyo Furitsu No. 1 Middle School formed the rather grandly named Japan Natural History Society (Nihon Hakubutsu Gakkai, est. 1900); within two years its membership had swelled to 159 young amateurs, who renamed the group Japan Natural History Confederation (Nihon Hakubutsugaku Dōshikai), a seemingly less academic title. The confederation undertook a plant collecting trip to mount Fuji in July 1901 and began publishing a magazine called *Hakubutsu no tomo* (Friends of natural history), which produced eighty-two issues between 1901 and 1911, after which it ceased production.[11] This group was a prominent example of how amateur plant aficionados banded together to explore their surroundings for species new to them—and sometimes new to academic biologists.

Makino Tomitarō, originally an amateur himself, guided many groups throughout Japan in plant collecting. From 1907 to 1912 he helped lead the Friends of Plants (Shokubutsu Dōkōkai, later renamed Makino Shokubutsu Dōkōkai) in the Tokyo-Yokohama region, meeting once each month on Sundays to discuss both wild and garden plants. Irrepressibly eager to spread public knowledge of plants, he traveled wherever invited in Japan to discuss the latest discoveries with interested citizens. He helped found amateur flower societies, even if members were more interested in ornamenting their gardens than in the science of plants. In ways reminiscent of the poet Matsuo Bashō (1644–1694), who traveled the hinterlands to bring the latest poetic techniques from Edo to local writers, Makino visited botanical clubs throughout the country, pioneering the creation of plant survey teams (*kansatsukai*) and encouraging members to catalog and publish their collections. Much of the regional botanical investigation was done by residents in their spare time; volunteers put out occasional newsletters after leading treks in nearby meadows and hillsides to identify rare or unusual species. The resulting local records—a kind of gazetteer of local flora—remain valuable to researchers and cram the shelves of major bookstores today. Those from Shiga, Yamaguchi, and Ishikawa prefectures are considered especially thorough. The rarest species gathered by clubs often ended up in the herbarium of Tokyo Imperial University.

In this way Makino's encouragement of public botany led to more discoveries of new plants by amateurs than by biologists.[12] Public botany also led to engaging women as well as men in the production of new scientific knowledge of nature.[13] A major vehicle of spreading public botanical knowledge was the magazine Makino founded in 1916, *Journal of Japanese Botany* (*Shokubutsu kenkyū zasshi*), which now as then combines scholarly rigor and accessible outreach. A cognate effect of Makino's efforts was the creation in 1942 of the Friends of Plant Taxonomy (Shokubutsu Bunrui Dōshikai, renamed Nihon Shokubutsu Bunrui Gakkai in 1948), led by researchers at Tokyo Imperial University with local branches throughout Japan and producing regular publications in plant systematics.[14]

Makino first named a plant species in 1889, Yamatogusa (*Theligonum japonicum, Cynocrambe japonica*), a perennial herb with a foul odor, followed by naming nearly 1,600 species of wild and cultivated plants throughout his long career. Because he didn't use Latin scientific names, the terms he applied after 1935 are considered nonstandard, including those in his hugely popular *Makino Nihon shokubutsu zukan* (1940). Most of Makino's names are still in use, including those of more than 300 Japanese vascular plants.[15] The Makino Herbarium at Tokyo Metropolitan University contains among the 400,000 items in its collection many of the rarest specimens he gathered, some of them still pressed between sheets of newsprint. The Kōchi Prefectural Makino Botanical Garden features his personal library of 45,000 books, as well as drawings and personal papers, cataloged and donated by his family. The prefectural garden itself is a kind of field museum meant to promote further research.[16] The Makino Memorial Garden and Museum at Higashi Ōizumi, Tokyo, was established by Nerima ward at the house where he lived starting in 1926. This small but handsome site contains a restful collection of trees that shade the residence through careful placement. At the ruins of the former Ikenaga Botanical Research Institute in Kobe is also a small park honoring Makino's memory.[17] No other Japanese scientist is recognized with such an array of memorials. Yet the best sign of the esteem now accorded Makino is that he has been the subject of two major exhibits organized by the National Museum of Nature and Science in Ueno park, Tokyo. The 1998 show honored his insights into the "spirit of vegetation"; the 2013 exhibit marked 150 years since his birth and drew an astonishing 860,000 visitors.[18] Kadota Yūichi (b. 1949), a scholar in the museum's botany department, summarized his high regard for Makino in 2012: "He became a central pillar of Japanese botany after arising from scratch. He wanted to make all Japanese plants known to the world. He worked unusually hard and exhibited powers of concentration unimaginable by ordinary persons. When you look at his uncompromisingly detailed plant drawings, you can understand this."[19] Makino set high stan-

Figure 5.1 Makino Memorial Garden and Museum in Higashi Ōizumi, Tokyo, contains an impressive sampling of trees in a compact space and reveals the simple surroundings where Makino Tomitarō spent his later years.

dards for himself and for the world of plant biologists, yet the throngs who attended the 2013 exhibit suggest that he gave the public broad access to botanical knowledge and stimulated interest in plant life that continues today.

Minakata Kumagusu too was honored with an exhibit at the science museum in 2017–2018 marking the 150th anniversary of his birth, titled "Minakata Kumagusu: An Informant-Savant 100 Years Ahead of His Time" (*Minakata Kumagusu: Hyakunen hayakatta chi no hito*). Mainly a self-taught naturalist and linguist who knew a dozen or more languages, Minakata embraced Mahayana Buddhism, especially Shingon esotericism, as a way of transcending Western science. Sociologist Tsurumi Kazuko has pointed out that Minakata regarded chance as key to understanding evolution, not necessity as Darwin argued. Minakata was also skeptical of Linnaean taxonomic classifications.[20] He regarded the mandala as useful for portraying connections between physical and mental realms, between consciousness and subconsciousness, whereas modern science was concerned only with tangible phenomena (he wrote in an age before modern psychology emerged as a full-fledged discipline).[21] Minakata sought "all wisdom" or to be "a person who knows everything" (*issaichi*),[22] a comprehensive epistemology in defiance of the increasing specialization of modern knowledge. Just as Buddhist holism accommodated both spiritual and material experience, Minakata studied the entire natural environment, of which people were a part. These convictions led him to explore how to link his botanical findings with the realm of the spirits and later to probe Japanese folklore,[23] a field that Minakata, Yanagita Kunio (1875–1962), Origuchi Shinobu (1887–1953), and others developed in the early twentieth century.

Minakata grew up in Wakayama prefecture, learned natural history in middle school, and spent the years 1883 through 1885 at University of Tokyo Preparatory School (Tokyo Daigaku Yobimon), where future novelist Natsume Sōseki (1867–1916) was a classmate. Minakata flunked out of school and moved to San Francisco in 1886, enrolling briefly at Pacific Business College, then equally briefly at Michigan State Agricultural College in East Lansing. For the next two years he was an independent student in Ann Arbor, reading Spencer and Darwin, collecting plants, and visiting museums at the University of Michigan. Minakata then traveled to Florida and several Caribbean countries to gather slime molds and mycological specimens before moving to London in 1892. Aided by a prodigious memory, he studied folklore and microbiology in the British Museum, compiling 10,000 notebook pages in 52 volumes called "London Extracts," and composing the first of the 50 articles and 323 notes and queries he contributed to *Nature* between 1893 and 1933.[24] His parents finally cut off their financial support in 1899, forcing Minakata to return to Wakayama in 1900.

For the next several years Minakata studied ferns, mosses, mushrooms, slime molds, algae, nonflowering plants, forests, and rock formations at Nachi and Tanabe in southern Wakayama. Like most naturalists, he was fascinated by obscure and unknown species, regardless of any practical use for medicine, foodstuffs, manufactures, or ornamentation. The specimens he amassed at Tanabe, although not considered especially sophisticated by today's standards, reflected this attraction to odd curiosities of nature as well as to more common varieties of plants. Hosoya Tsuyoshi (b. 1963) of the botany department at the National Museum of Nature and Science has observed that "from today's standpoint, we would hardly call him a research scientist" who proposed hypotheses and tested them. Hosoya considers Minakata a naturalist with strong views, few of whose papers were written with scholarly rigor.[25] What he did accomplish, however, was to describe the ecology of the entire southern Wakayama region, becoming one of the first specialists to use the term *seitaigaku* for ecology.

Minakata considered slime molds (any of several varieties of primitive eukaryotic organisms), which earlier had been classed as fungi, to exist somewhere between plants and animals and to be elemental to the origins of animal life. Of the 1,000 known species of slime molds in the world, about 600 are found in Japan. As of 1927, when he stopped collecting, Minakata had identified 196 species and a new genus he discovered in his garden, *Minakatalla longifila*. Zeal for his research led him to write Crown Prince Hirohito in 1926 that slime molds were protozoa, the elementary forms of animals rather than plants. Three years later he lectured Hirohito, who by then was the Shōwa emperor, on the subject.[26] Perhaps prompted by Minakata's fervor for studying slime molds, Hirohito, who expressed a lifelong interest in plant biology, began research at Shimoda in 1972 and for a time specialized in myxomycetes, a class of slime molds.[27] Today slime molds are regarded as neither plants nor animals, leaving open the questions raised by Minakata a century or more ago. Minakata also collected many of the 1,600 species of freshwater and marine algae known to exist in Japan, as well as gathering more than 700 specimens of lichens, mostly unidentified because no reference work on lichens was available in Japan at the time. Minakata likewise collected about 6,000 species of fungi, or half the total thought to exist in Japan in his day, and he drew roughly 4,000 full-size watercolor illustrations of the fungi in his possession.[28] He also prepared about 4,000 descriptions of seaweeds and 7,000 specimens of slime molds.

From 1906 to 1912 the Japanese government, pressed for funds after the costly war with Russia and eager to exert greater control over localities, carried out a movement to consolidate Shinto shrines, forcing village shrines to merge with those in nearby towns and cities. As a result, the number of shrines with official status (and state subventions) fell from 5,853 to 463—a drastic change in villagers' sense of place, worship, and festival. Most former shrines became property of local governments, many of which promptly sold the surrounding shrine forests for logging. Minakata was furious at seeing forestlands in his region disappearing, so he peppered local newspapers with letters of protest arguing that the loss of trees meant the extinction of forest-floor vegetation as well as loss of animal habitats. In August 1910 he was arrested for trespassing when he hurled a bag of specimens in a meeting at Tanabe Middle School after being refused permission to speak with a government official promoting the shrine mergers to local residents in attendance. He spent eighteen days fuming in jail.[29] Minakata intensified his campaign in 1911, addressing two letters of protest to Matsumura Jinzō at Tokyo Imperial University, the doyen of establishment botanists, in which he posited spiritual as well as ecological harm if the forest cutting continued. He worried that destroying village shrines would create spaces for Christianity and new Japanese religious sects to enter. Minakata entrusted the letters to folklorist Yanagita Kunio, then an employee of the cabinet legislation bureau, for delivery to Matsumura.

Instead Yanagita published them as *Minakata nisho* (Two letters from Minakata) and distributed fifty copies to experts in various fields, seeking their support for Minakata's views. The result of these efforts was meager: a few valuable trees were spared, and one or two shrine forests were left untouched along the Kumano Kodō pilgrimage route near his home. The shrine consolidation program was abolished by parliamentary action in 1920.[30]

Of his many writings on plant biology, Minakata's manuscript *Minakata Kumagusu kinrui zufu* (Colored illustrations of fungi), eventually published by Shinchōsha in 2007, distilled his painstaking research on mushrooms and other fungi. In his later years Minakata turned increasingly to folklore, Buddhist thought, and movements to protect the natural environment in his region of Wakayama prefecture. The campaign to fund a Minakata Botanical Institute at his residence in Tanabe fizzled, but today his house is a handsome archive containing his manuscripts, drawings, and extensive library. In some ways Minakata's career returned the study of plants in Japan full circle to the natural history practiced on the eve of modern plant science in the mid-nineteenth century. But in other ways his international experience, unparalleled among the leading Japanese botanists of his era, together with his lack of an academic position that might impose excessive specialization, gave him license to research more broadly, in more forms of plant life, than any other scholar in his lifetime. His diverse interests in myths, folklore, Buddhism, philosophy, ethnology, and ecological activism should not mask his achievements in identifying rare species and opening up new fields of botanical research in Japan. In this sense his career foreshadowed the wide-ranging investigations of the next generation of plant biologists, both those in Japan and those for whom their field laboratory became the entire Japanese empire.

Japanese Plant Biology in the Early Twentieth Century

The first decades of the twentieth century in Japan were a time of genera- or species-specific research by young scholars who built their careers on the broad foundations of botanical knowledge provided by Makino, Minakata, and their predecessors. Significant studies focused on such topics as male ferns, fungi, hostas, lichens, and Saint John's wort, among others. In 1926 and shortly thereafter, eleven monographs appeared on single plant genera. Other scholars, often assisted by public botany clubs, produced carefully defined studies of localities such as Karuizawa, the Ogasawara islands, and Matsushima bay.[31] For example, Hayata Bunzō (1874–1934), who spent nineteen years researching Taiwanese plants and directed the Koishikawa Botanical Gardens from 1924 to 1930, published his findings about plants on mount Fuji (altitude 3,776 meters) in 1911 and 1926. Hayata cataloged 739 species of 96 plant families on the mountain and calculated that its vegetated 1,295 square kilometers included 52 percent grasses, 11 percent evergreen conifers, and just 6 percent alpine plants.[32] He also emphasized *iyachi* (undesirable land), the tendency of plants to come to dislike a place across time, leading to their replacement by other species.[33] Ecologists generally call this process ecological succession.

At the same time several general works were issued to mark the progress in Japanese plant science since the Meiji period. Nemoto Kanji (1860–1936) and Makino Tomitarō published *Nihon shokubutsu sōkan* (Overview of Japanese plants, 1925), which updated Matsumura Jinzō's work in *Teikoku shokubutsu meikan* (Directory of plants in the empire, 2 vols., 1904–1912).[34] The findings recorded in these and other synoptic books were further revised and abbreviated in *Nihon shokubutsu meii* (Names of Japanese plants, 1939), a 201-page handbook edited by Honda Masaji (1897–1984). At the same time Honda and Nakai Takenoshin produced nine volumes of a comprehensive illustrated work, *Dai Nihon shokubutsushi* (Greater Japanese plants, 1938–1943), that was curtailed before completion because of wartime paper shortages. Other abundantly illustrated books appeared in the 1920s and 1930s, many of which were reprinted after World War Two, but none had the mix of scholarly and commercial appeal of *Makino Nihon shokubutsu zukan* (1940).[35]

In something of a reprise of practices from the 1870s and 1880s, Professor Asahina Yasuhiko (1881–1975) of the Tokyo Imperial University medical school sent lichen specimens to Alexander Zahlbruckner (1860–1938) in Vienna for authentication, but later he ceased relying on foreign authority and did his own taxonomy and analysis. In the 1930s Asahina developed a microchemical detection method for determining the components of lichens, a process easy for taxonomists to use that became the standard detection method for the next fifty years. Asahina was a leading light in chemotaxonomy and ended up identifying new genera of mosses.[36] Another inventive scholar was Shibata Keita (1877–1949), who in 1931 published Japan's first monograph on photosynthesis, *Tanso oyobi chisso no dōka sayō* (Carbon and nitrogen assimilation). Shibata argued that the allotrope of oxygen (O_2) evolved in photosynthesis was entirely derived from water, several years before the discovery of the light-dependent Hill reaction of photosynthesis in 1937.[37]

Ogura Yuzuru (1895–1981), a specialist on systematic morphology, studied fossils, including pteridophytes (also called cryptogams)—ferns, horsetails, and lycophytes that reproduce via spores, not seeds or flowers. His *Shokubutsu keitaigaku* (Plant morphology, 1934) became an essential guide for that field in Japan. Fujii Kenjirō (1866–1952), the plant geneticist who specialized for a time in ginkgos, also strengthened the basis for morphology and was among the first in Japan to bridge cytology and morphology through his research on

Sigillaria, a genus of extinct, spore-bearing tree-like plants.[38] Fujii's student Tahara Masato (1884–1969), a cytologist at Tohoku Imperial University, gained fame for discovering multiple genera of chrysanthemums in 1915. Tahara was an unusual polymath in an age of specialization whose most important research included the cytology and embryology of both angiosperms and gymnosperms, as well as the morphology of seaweeds. Another Fujii student, Kuwata Yoshinari (1882–1981), specialized in plant chromosomes during a lengthy career at Kyoto Imperial University. His research team contributed many studies of chromosomes in nuclear fission, such as the movement of chromosomes and hydration and evaporation. This work drew Cyril Dean Darlington (1903–1981), an English geneticist and eugenicist, and a number of other specialists from abroad to join Japanese colleagues in research on chromosomes.[39] In Hokkaido the plant geneticist Matsuura Hajime (1900–1990) became an expert on the genus *Trillium,* a perennial herbaceous plant growing from rhizomes. Matsuura was well known internationally for his 787-page *A Bibliographical Monograph on Plant Genetics (Genic Analysis), 1900–1925,* published in 1929. After the war he got into a scrape with American occupation authorities when, during a nationwide purge of communists in 1950, a speaker from GHQ was shouted down by students at Hokkaido University. As dean of the science faculty he was forced to take responsibility for the incident.[40]

Despite the pathbreaking work of Miyoshi Manabu and Minakata Kumagusu, the field of ecology proved difficult to establish in Japan before World War Two. As the official history of the Botanical Society of Japan (1982) pointed out, ecology was for many years considered a part of physiology. Someone impatient with this subordinate classification was Nakano Harufusa (1883–1973), a student of Miyoshi's who specialized in green algae but also argued vociferously that ecologists should focus on plant communities such as groups of trees. Appointed as a professor at Tokyo Imperial University in 1924, he succeeded in setting up a plant ecology laboratory in 1934, the same year he was promoted to professor. His major publication was the 573-page *Shokubutsu seiri oyobi seitaigaku jikkenhō* (Experimental methods in plant physiology and ecology, 1933).[41] Another student of Miyoshi's was Yoshii Yoshiji (Yoshitsugu, 1888–1977), who was appointed to Japan's first chair in ecology at Tohoku Imperial University in 1928 and used his position as a platform to promote plant ecology throughout the country. The following year he became the founding director of the university's Mount Hakkōda Botanical Laboratory (Hakkōdasan Shokubutsu Jikkenjo), giving him an opportunity to oversee ecological research on forest plants. The laboratory began publishing a journal in 1935 called *Seitaigaku kenkyū* (continued as *Ecological Review* in 1959), the first of a number of ecology journals to appear in Japan. After the war Yoshii served as first president of the Plant Ecology Society (Shokubutsu Seitai Gakkai), founded in 1951 and subsumed by the Ecological Society of Japan (Nihon Seitai Gakkai), established at Tokyo's Institute for Nature Study (Shizen Kyōikuen) in 1954.[42] These and many other early twentieth-century scholars developed the study of plant life in Japan to a much greater level of empirical detail and methodological sophistication than their late nineteenth-century predecessors—but without whom their research would scarcely have been possible.

The Botany of Empire

Some Japanese historians believe that their country's first overseas colony was a territory in southern Korea known as Mimana (Kr. Imna), interposed between the early Korean states of Silla to the east and Baekje to the west, until Japanese settlers were evicted by Silla in the sixth century CE. The main source for the existence of Mimana is *Nihon shoki* (Chronicles of Japan, 720), an official court history dedicated to Empress Genshō (683–748, r. 715–724), but the events described in this work are disputed by scholars in both Japan and Korea, many of whom doubt that such a colony existed. But there is no uncertainty about a second expression of Japanese imperialism, the invasions of Korea in 1592 and 1597 ordered by Toyotomi Hideyoshi (1537–1598) to subdue both the Joseon dynasty in Korea and the Ming dynasty in China. After a prolonged stalemate and a renewed influx of troops, Japanese forces were driven out in 1598, and peace was restored to Korea until invasions from China in 1627 and 1636.

Korea was once again a target of Japanese imperialism in 1876, when the new Meiji government used gunboat diplomacy to force Joseon leaders to sign the Treaty of Ganghwa, opening the country to trade with Japan and granting Japanese settlers legal rights on Korean soil similar to those enjoyed by Westerners in Japan through unequal treaties imposed by the United States and European powers starting in 1858. The first Sino-Japanese war (1894–1895) was fought mainly in Korea and resulted in Japanese victory, freeing the Joseon dynasty from Chinese influence and giving Japan control over Taiwan in 1895, although not without fierce guerrilla resistance for two years before Japan established its colonial rule over the island, lasting until 1945. The Russo-Japanese war of 1904–1905 was contested mainly over Korea and Manchuria and resulted in Japan's surprising victory. The ensuing Portsmouth treaty awarded Japan the leasehold over Port Arthur (Dalian) in Manchuria; the southern half of Sakhalin island, which the Japanese colonized as Karafuto; and recognition as the dominant power in Korea, which Japan annexed as a colony in 1910. The formal Japanese empire was now complete territorially, but

Japanese imperial expansion through investment, trade, diplomacy, and military aggression continued for the following thirty-five years in Manchuria, China, Southeast Asia, and South Pacific islands until Japan's disastrous defeat in World War Two by Chinese, United States, British, and other Allied forces in August 1945.

Imperial botany and its more geographically defined cousin colonial botany trace their origins to the rise of large ships capable of transoceanic trade in early modern times. Wealth from world areas incorporated into European empires depended partly on the cultivation of profitable plants, both economic and ornamental. One influential study asserts that "early modern botany both facilitated and profited from colonialism and long-distance trade, and that the development of botany and Europe's commercial and territorial expansion are closely associated developments"[43]—a kind of political economy of plants. Some even called plants "green gold" because of their value to the imperial exchequer. In one reckoning, Europeans had founded 1,600 botanical gardens by 1800, the majority as agricultural experiment stations for cash crops and medicines.[44] Yet science and curiosity were equally central motives for imperial botany: early modern Spain emphasized knowing the empire by making the colonies visible through botanical images.[45] Elsewhere, as Christopher Parsons has written, "The creation of gardens in urban centers and at palaces was part of a broader strategy of political centralization and the development of a visual and material culture of royal authority in early modern France."[46]

Japan was a latecomer among modern empires, so the idea of colonial botany as understood in a European context offers only partial insights into the country's botanical practices in its overseas realm.[47] As Japanese control over East and Southeast Asia expanded, opportunities for botanical study and plant collection opened up, first in established colonies in Taiwan, Karafuto, and Korea, then in occupied areas such as Manchuria (northeast China, seized by Japan in 1931) and multiple countries in Southeast Asia overrun by Japanese military forces between 1941 and 1945. In evaluating the role of specialists on vegetation in Japan's overseas possessions, an alternative term more appropriate than *colonial botany* is *the botany of empire,* including not just formal colonies but also areas under military occupation. In these spaces researchers from Japan practiced a botany of collection rather than exploration, let alone domination. Especially popular were the numerous tropical plant gardens that sprang up in Japan to bring exotica to the mainland as emblems of Japan's imperium, yet the aims of scholars who set forth abroad were mainly scientific investigation and secondarily economic analysis, not imperial display. As early as 1903 the leading journal *The Botanical Magazine, Tokyo* began to turn outward with articles on Taiwanese plants and Korean ferns, launching a flood of scholarship on almost every corner of Japan's Greater East Asian Co-

Prosperity Sphere (est. 1940), a zone of economic self-sufficiency backed by military force covering East and Southeast Asia.

Long before Japanese forces invaded the region, Southeast Asia was a near-constant focus of plant collecting by Japanese researchers, as were islands in the Marianas and Carolines. Kanehira Ryōzō (1882–1948), who carried out detailed surveys in Taiwan and Southeast Asia, also became a leading expert on plant life in New Guinea, Micronesia, Melanesia, and Polynesia, a geographical reach matched by few if any other scholars.[48] His major work, written for the Japanese government's South Pacific Agency (Nan'yōchō), was the 468-page *Nan'yō guntō shokubutsushi* (Plants of the South Pacific Islands, 1933), a study of vegetation on Pacific islands mandated to Japan after World War One.[49] Further north Honda Masaji published a 484-page study of wild and cultivated grasses based on extensive research from Karafuto through the main Japanese islands to Taiwan, titled *Monographia Poacearum Japonicarum* (1930). This work identified 12 species new to Japan and tabulated 4 genera, 87 species, and 51 varieties of gramineous plants (*Poaceae,* a family of true grasses with 10,000 species worldwide).[50]

Deep in World War Two, Masamune Genkei (1899–1995), a professor of botany at Taihoku [Taibei] Imperial University, published *Kainantō shokubutsushi* (Plants of Hainan, 1944), with particular emphasis on forests and orchids. Another scholar with a Pan-Asian compass was Horikawa Yoshio (1902–1976), a specialist on mosses who collected specimens ranging from Karafuto and the Kuril islands to Ogasawara, Ryukyu, Korea, Taiwan, and parts of China. Most of his collection was lost in the Hiroshima atomic bombing in August 1945, so after Japan's surrender he collected fresh samples "astronomical in number" within the nation but not overseas because of postwar travel restrictions. Horikawa's astonishing persistence in collecting specimens destroyed in wartime resembled that of lexicographer Morohashi Tetsuji (1883–1987), editor of the thirteen-volume *Dai Kan Wa jiten* (Great Chinese-Japanese dictionary, 1956–1960). Morohashi spent twenty years researching Chinese texts, only to have his manuscript incinerated in an American air raid on Tokyo after just one volume had been published. Undaunted, following the war he redid his research on the same texts, but this time it took him just ten years. In a similar fashion, by 1955 Horikawa had rebuilt his collection by amassing fifty thousand specimens of mosses, hornworts, and liverworts and twenty thousand specimens of other vascular plants.[51]

Unlike Kanehira and Horikawa, most botanists of empire focused on single overseas territories. Karafuto became a Japanese possession under the Portsmouth Treaty of 1905 and a formal colony two years later. The treaty awarded Japan the southern half of Russia's Sakhalin island below 50 degrees north latitude, a recognition that Japanese had settled parts of the island in the Edo period as Kita Ezo (northern Hokkaido) and had unsuccessfully laid claim to all of Sakhalin in 1845. The sparsely populated colony chiefly provided Japan with primary products such as timber, coal, and fish; efforts to establish permanent agricultural settlements largely failed.[52] The first Japanese botanical foray to Karafuto came in 1906, when elementary schoolteacher Nakahara Genji (fl. early twentieth century) collected 300 species of higher plants during a three-month visit, after which he sought help classifying them from Koizumi Gen'ichi (1883–1953), a student at Tokyo Imperial University who later became the chief botany professor at Kyoto Imperial University. Koizumi published the results of Nakahara's collecting in the *Memoirs* (*Kiyō*) of Tokyo Imperial University's College of Science in 1910.

The most comprehensive survey of plants in Karafuto came with the publication of the four-volume *Karafuto shokubutsu zushi* (Illustrated plants of Karafuto, 1937–1940) by Sugawara Shigezō (1876–1967), who conducted extensive tabulations of vegetation starting in 1922 for the Karafuto administrative agency. His companion *Plants of Saghalien* (1937) provided a history of writings on the botany of the island from 1753 to 1936, listing 137 publications on the flora of Sakhalin and nearby subarctic regions.[53] Sugawara's son Samukawa Kōtarō (1908–1977), a novelist who grew up in Karafuto, described the colony as "the terminus of Japan,"[54] or northernmost outpost of empire, which partly explains the fascination of plant biologists with this remote territory. Research on marine plants in Karafuto was carried out starting in the 1920s by Tokita Jun (1903–1990), a student of Miyabe Kingo's at Hokkaido Imperial University. In 1940 Tokita was appointed assistant professor of marine botany at the university, and five years later he completed his major study of cold-water oceanic plants, *The Marine Algae of Southern Saghalien*.[55] These and other researchers provided detailed censuses and illustrations of plant life in the colony, almost entirely for scientific purposes because so few species on the island added value to the metropole.

Taiwan was the opposite, a colony of great economic worth to imperial Japan within a half dozen years of its transfer to Japanese administration under the 1895 Treaty of Shimonoseki. The Japanese colonial government-general (Sōtokufu) included within its home affairs bureau an office of agriculture, responsible for land surveys, extension services, and economic development of farm crops for the metropole's benefit. From 1900 to 1920 sugar was the main export to Japan, replaced by rice thereafter.[56] Forest products were also sent to the metropole: reflecting the imperial symbolism of Tokyo's Meiji shrine (opened 1920), the two great torii gateways to the shrine complex were erected using ancient cypress from Taiwan. When Taiwan began to industrialize in support of Japan's military expansion in the 1930s, the island's agricultural surpluses helped feed the colony's urban working class and generate

Figure 5.2 Tokyo's Meiji Shrine opened in 1920 with a great torii gateway of cypress from Japan's Taiwan colony, 2009.

land taxes for industrial investment. Botanists from the Japanese mainland included economic plants among their studies during the early years of colonization but turned more and more to scientific surveys in the 1930s and early 1940s, building a formidable budget of knowledge about the impressive greenery found everywhere on the island.

Taiwan was a land of considerable botanical abundance when Japan incorporated the island into its empire. An 1896 survey by Irish plant collector Augustine Henry (1857–1930) identified 1,428 varieties of land plants in Taiwan, as well as a few seaweeds.[57] That same year Tokyo Imperial University, after years of neglecting botanical studies of regions beyond Japan, established a Taiwan research group, dispatching Makino Tomitarō and others to the colony to investigate its vegetation. One result was the first major study of lowland plant life on the island, *Taiwan shokubutsu meii* (Names of Taiwanese plants, 1906) by Matsumura Jinzō and Hayata Bunzō.[58] Hayata followed this book with *Flora Montana Formosa* (1908), enumerating 392 species in 266 genera he found at higher elevations; he reached the uncertain conclusion that these plants generally resembled those in Japan more than China.[59]

Hayata was Matsumura's student at Tokyo Imperial University, and after spending the years 1905 through 1924 carrying out plant inventories for the government-general in Taihoku, he returned to the university as professor of botany in charge of taxonomy and director of the Koishikawa garden. As part of his duties for the colonial administration, he traveled to London for research on Taiwanese plants at the Royal Botanic Gardens, Kew, followed by visits to France, Germany, and Russia in 1910. These travels made him one of the first Japanese taxonomists to be known personally in the wider world of scholarship.[60] The government-general sent him to French Indochina in 1917 and 1921 to supervise plant inventories, followed by visits to Thailand, Indonesia, and China's Yunnan province. Among the many reports and plant censuses he prepared for the colonial administration were studies of forest products, crops and other plants of economic importance, and tropical vegetation. His scientific interests included hereditary characteristics and lineages of ferns; he became famous for advocating novel approaches to botanical classifications. These were evident in his multivolume *Taiwan shokubutsu zufu* (Illustrated Taiwanese plants, 1911–1921), which tabulated more than 2,000 plants and built the foundation for further research there. Hayata brought large numbers of herbarium specimens to Tokyo Imperial University when he returned as a professor in 1924. He suffered a heart attack shortly after New Year's Day 1934, briefly taught his classes at home, and died on January 13 of that year.[61]

Two of Hayata's prominent students at the university, Masamune Genkei (1899–1995) and Yamamoto Yoshimatsu (1893–1947), became professors of botany at Taihoku Imperial University (today National Taiwan University) shortly after the institution was founded in 1928. Yamamoto was the only student of Taiwanese botany when he entered Tokyo Imperial University in 1920; his appointment at Taihoku Imperial University in 1928 allowed him much scope to collect plants throughout Southeast Asia, where he identified about 180 new species. In a sign of the relative amity between Japanese and Taiwanese scholars, after Japan's defeat and the end of its colonial administration in 1945 Yamamoto was named a professor at National Taiwan University, but he succumbed to typhus and died in 1947.[62] For his part, Masamune in 1934 published the 637-page *Floristic and Geobotanical Studies on the Island of Yakushima, Province Osumi* (an island south of Kagoshima prefecture), enumerating 1,143 species, varieties, and forms of plants and identifying broadleaf evergreens on Mesozoic slate, while conifers predominated on weathered granite at higher altitudes.[63] This detailed study won Masamune wide respect and served as an introit to his subsequent research on Taiwan and Hainan.

In Taiwan, Korea, and elsewhere in Japan's imperial realm botanical gardens and public parks were tools of colonization of occupied peoples by authorities bent on social management. In Taiwan, and to a degree in Korea, such green spaces were spatial levers of Japan's empire right down to 1945. Taibei Nursery Garden, opened in 1900 and renamed Taibei Botanical Garden in 1912, served Japan's colonial strategy of acculturation by admitting the public to visit its compact but lush grounds and minizoo.[64] The garden's juxtaposition of plant specimens from throughout Japan's empire suggested cultural contiguity but not yet hybridity—the latter a goal the Japanese sought largely in vain both in Taiwan and in Korea during World War Two. Taiwan was more prosperous but politically and industrially less prominent than Korea during the colonial period; its botanical specimens were far less familiar to people in Japan than Korea's vegetation, accounting for the Japanese fascination with tropical plants from afar that continues to this day. In this sense, Taiwan was a site of plants in motion, both inbound for display at the Taibei Botanical Garden and outbound for the delectation and subtle imperialization of the Japanese public.

Korea, which was twice as populous as Taiwan and geographically much closer to Japan, was deemed hugely important to the Japanese empire from the moment of annexation in 1910. By that year as many as 170,000 Japanese sought economic opportunity in Korea, especially farmland; in 1932 a majority of arable land was thought to be owned by Japanese, although with what degree of actual control is uncertain.[65] The Japanese colonial government-general in Keijō (Seoul) pursued a policy of mercantilism to benefit the metropole. In seeking economic integration with Korea, Japan developed primary education to train children for the workforce, constructed transportation networks, promoted public health measures, and built infrastructure that after 1945 formed the foundation for South Korea's economic growth under President Park Chung-hee from 1961 to 1979. To a greater degree than in Taiwan, colonial administrators exploited Korea's coal, iron ore, lumber, and marine resources for Japan's benefit; they also sought to increase rice output for export to the metropole. In the 1930s and early 1940s Japanese encouraged industrialization in chemicals, steel, and munitions, especially in the northern part of the peninsula. Yet the consensus is that Japan's colony in Korea never earned the metropole a profit, whereas Taiwan proved profitable within a half dozen years of Japan's takeover.

Nakai Takenoshin, who studied under Matsumura Jinzō and became a professor of taxonomy at Tokyo Imperial University starting in 1922, trained a generation of specialists on the flora of Korea while also directing the Koishikawa Botanical Gardens from 1930 until his retirement. Nakai made eighteen collecting trips to Korea between 1909 and 1940, focusing on mountains, and he eventually enumerated 3,176 plant species in the colony. He published more than 500 botanical papers and proposed 4,000 new names of Japanese and Korean plants. Nakai's *Flora Koreana* (2 vols., 1909–1911) meticulously listed scholarly articles about each plant included in the book.[66] Especially influential were his 455-page *Chōsen shokubutsu* (Korean plants, 1914) and *Chōsen shinrin shokubutsuhen* (Forest plants of Korea, 5 vols., 1927–1932), the latter being one of the many works he wrote for the government-general in Keijō.[67] In this way Nakai balanced the instrumental needs of the colonial administration with the imperatives of scientific inquiry throughout his career. Horticulturalist and garden administrator Uchiyama Tomijirō (1851–1915) and botanist Hotta Teikichi (1899–1976), a specialist on mulberries, were two of the numerous Japanese investigators who followed Nakai's path in doing scientific research in colonial Korea.[68] Their work contrasted with self-styled settler experts such as the forester Ishidoya Tsutomu (1891–1958), who tried but could not overcome the asymmetries of authority as a Japanese trying to collaborate with colonized botanists in Korea.[69] Hayata, Nakai, and other Japanese plant biologists went to the colonies as scholars in the service of science as well as the state; their reports were technical, descriptive, and data-driven, shorn of nearly all admiring language about the gorgeous landscapes in which they worked. Korea's natural environment was equal in beauty to Taiwan's but larger in scale, providing abundant sites for tourism from Japan, so that the handsome scenery each colony offered visitors needed no acclaim from plant specialists.[70] The task of Japan's botanists of empire was collection and analysis, not exploration and discovery as performed a century earlier in East Asia by Europeans and Americans, who often narrated their adventures in lyrical language for the benefit of armchair travelers back home.

The education ministry sent Nakai to visit important herbaria in Europe and the United States between 1923 and 1925; there he learned that Japan's level of plant taxonomy "wasn't different from Europe and America."[71] Like Makino Tomitarō, Nakai was a fervent supporter of public botany, praising local amateurs for their discerning eyes when they sent him rare specimens from all regions of Japan. Honda Masaji succeeded Nakai at Tokyo Imperial University and likewise encouraged amateur botanists.[72] Nakai's career took a further progovernment turn in 1942 when he served the occupying Japanese military administration in Indonesia as chief of a botanical garden in Java. He was named head of the Tokyo Science Museum in 1947 and continued as director when the facility was renamed National Science Museum in 1949 (today National Museum of Nature and Science).[73]

As with Taiwan and Korea, Manchuria was a locus of study by Japanese plant collectors as early as the 1880s. In 1904 Yabe Yoshisada (1876–1931) began teaching at Peking University and soon undertook research on plants in the region. His *Minami Manshū shokubutsu mokuroku* (Catalog of plants in southern Manchuria, 1912) established a baseline for further botanical research by younger scholars such as the seaweed specialist Noda Mitsuzō (1909–1995), who did field work in Manchuria after it was renamed Manchukuo (Manzhouguo) and became an informal Japanese colony in 1932. Japan installed the last Qing emperor, Xuantong (Aisin Gioro Puyi, known as Henry Pu Yi, 1906–1967) as head of the nominally independent state of Manchukuo; imprisoned by the People's Republic for a decade, Pu Yi spent his final years as a gardener at the Beijing Botanical Gardens.[74] Noda visited Manchuria at various times during the second Sino-Japanese war (1937–1945) and summarized his findings in *Chūgoku Tōhokuku (Manshū) no shokubutsushi* (Flora of the northeast province [Manchuria] of China, 1971). Noda's near-contemporary Kitagawa Masao (1910–1995) was born in Dalian and studied botany at Tokyo Imperial University, specializing in ferns, before returning to Manchuria to survey plant life as an employee of the Japanese government's Institute of Scientific Research, Manchukuo (Tairiku Kagakuin Kenkyūkan). In 1933 Kitagawa, Nakai Takenoshin, and Honda Masaji helped establish the quasiofficial Manchuria-Mongolia Scholarly Survey Research Group (Manmō Gakujutsu Chōsa Kenkyūdan) to investigate natural history, archaeology, ethnology, and related topics in Manchukuo and Inner Mongolia.[75] Most of the herbarium specimens Kitagawa sent to his alma mater were lost when an American air raid at the end of World War Two damaged Science Building No. 2 at the Hongō campus. Unlike Noda, Kitagawa was able to publish his scientific findings during wartime, producing his 487-page *Lineamenta Florae Manshuricae, or an Enumeration of All the Indigenous Vascular Plants Hitherto Known from Manchurian Empire,* published in 1939 at Hsinaking (Changchun), the capital of Manchukuo.[76]

Yabe, Noda, Kitagawa, and most other botanical researchers in Manchuria depended on some degree of support from the semipublic South Manchuria Railway Company, established by Japan in 1906 to operate portions of the former Chinese Eastern Railway as a gateway to exploiting the mineral and industrial resources of the region. Known as Mantetsu, the company published Yabe's 1912 plant catalog and numerous other studies of economic and political life in both Manchuria and north China. Mantetsu operated an agricultural experiment station at Gongzhuling (Jilin province), later the site of a Japanese air base. The Gongzhuling station produced two varieties of dry upland rice as early as 1922; other Mantetsu researchers focused on soy oils and soy meal. In short, the company had its fingers in most botanical research, both instrumental and scientific, in the region until it was dissolved in 1945. Its research archives remain indispensable for scholars of northeast and north China today.[77]

During World War Two senior Japanese researchers such as Nakai Takenoshin and Honda Masaji were dispatched to Penang, Singapore, New Guinea, and elsewhere in Southeast Asia to survey tropical plants, working with local assistants under difficult conditions. Once liberated by Japan's surrender in 1945, Japanese plant biologists extended their studies to India, Nepal, the Himalaya, Tibet, Afghanistan, Karakorum, and the Andes. At the same time, many of Japan's overseas specialists brought the botany of empire home by setting up laboratories at established campuses as well as new universities and research institutes throughout the country. Prominent examples include Masamune Genkei, who moved to Kanazawa University to focus on plants along the Japan sea coastline; Kitagawa Masao, who was appointed at Yokohama National University; and Noda Mitsuzō, who accepted a post at Niigata University. Nakai Takenoshin became head of the National Science Museum in Ueno park, while Sugawara Shigezō moved from Karafuto to Hokkaido Gakugei University.[78] Despite inadequate funding and the vicissitudes of economic rebuilding between 1945 and 1955, Japanese interest in the botany of Asia flourished after the war and remains vigorous at such major centers as Kyoto University and the Tsukuba Botanical Garden. In this sense, the end of the politico-military empire meant a fresh start for scientific investigations of Japan and its regional botanical environment in East and Southeast Asia. This postwar context forms the framework for the chapter that follows.

Six

Plant Biologists in an Era of Specialization

"To understand Japanese doctoral training in science, you have to understand that students start to become motivated only when they have to write a research thesis in their fourth undergraduate year. When they enter graduate school, they don't take courses for a broad knowledge of the discipline; instead they specialize early and stick with their specialty the rest of their career."[1] With these words Azuma Yoshirō, a highly respected professor of physics at Tokyo's Sophia University, pinpoints a long-standing aspect of scholarly training in Japan that grew even more conspicuous after 1945 compared with practices in other countries: "Graduate students serve an apprenticeship to their professor rather than following an organized curriculum as in the United States." In Azuma's view, most science education is based on assimilating the received wisdom of the field; graduate students "engage in critical inquiry and debate only at advanced stages of their training, including as postdocs."[2] This combination of absorbing past knowledge and scant critical challenge of established ideas sets stiff limits on exploration and innovation in Japanese botany no less than in other areas of the physical and life sciences. Like generations of critics before him, Iwatsuki Kunio, a distinguished biologist and professor emeritus at the University of Tokyo, adds that universities should abolish the system of professorial chairs (*kōza*), whereby retiring professors choose their successors from among their acolytes; ending the system would allow universities to introduce fresh topics to the curriculum and research profile.[3] Acknowledging lacunae in how Japan prepares its scientific researchers, other scholars nonetheless hold that training for careers in the sciences is merely different from overseas, not necessarily better or worse.

Although a number of Japanese researchers earned international recognition from fellow botanists before World War Two, through the year 2018 no scientist whose research included plants and plant-based materials appears on the list of Japanese Nobel prizewinners in chemistry, medicine, or physiology (there is no prize specifically in biology). Nine Japanese citizens have won Nobel prizes in physics, 7 in chemistry, 5 in medicine or physiology, 2 in literature, and 1 in peace; 2 other winners in physics were born in Japan but became United States citizens, and Kazuo Ishiguro, the 2017 laureate in literature, was born in Nagasaki but became a British subject in 1983. Two Japanese have been nominees for Nobel prizes (but not winners) based on their research in plant sciences, Suzuki Umetarō (1874–1943) and Asahina Yasuhiko (1881–1975), both professors at the University of Tokyo. Suzuki, nominated in 1914 (medicine and physiology) and 1936 (chemistry), was noted for studies of rice bran leading to identification of thiamine or vitamin B one,[4] while Asahina was nominated in 1951 and 1952 in chemistry for his research on lichens.[5] Even the International Prize in Biology, awarded since 1985 by the Japan Society for the Promotion of Science (Nihon Gakujutsu Shinkōkai), has never been conferred on a Japanese researcher in plant sciences, although seven Japan-born scholars have won this prize in other areas of biology. The lack of international recognition accorded Japanese botanists can be explained partly by the conventional training outlined by Azuma Yoshirō but also because research on plants has less visibility and is "not perceived as having a direct bearing on human health"[6]—an obviously myopic misperception, given the essential place of plant-based foods in the human diet.[7]

As with scientific research everywhere, the key themes of postwar botanical research in Japan were specialization, diversification of inquiry, new infrastructure, innovations in technology, and the emergence of ecological perspectives. This chapter examines the intense preoccupation with Japanese plant life by scholars in that country in the later twentieth century, as well as the intellectual and material contexts within which they carried out their investigations. Today the plant sciences in Japan are robust, and new discoveries emerge regularly, after overcoming a slow and impoverished recovery from wartime desiccation during the 1950s.

Enhanced Infrastructure for Research

The institutional foundations for plant biology in Japan, as in many other countries, are state-supported universities, botanical gardens and arboretums, scholarly societies, and government funding. Since the 1870s innovative Japanese research in the plant sciences has been mainly concentrated in faculties of science and agriculture at state-supported universities or, in the case of medicinal plants, in private gardens maintained by pharmaceutical firms. Under the new university system promulgated in 1949, Japan's prewar higher schools, vocational schools, and teacher-education schools were mostly combined by each prefecture or else joined with existing universities to form new universities. In addition to the seven former imperial universities within Japan, sixty-six universities soon provided some form of botanical research and instruction. By 1955 nine of the new universities had created separate faculties of science (*rigakubu*), eventually offering M.Sc. degrees in various fields, while Tokyo Metropolitan University (1949), Osaka City University (1960), and Kobe University (1981) also established new Ph.D. programs in the sciences. Aided by revenues accruing in an era of high-speed economic growth in the 1960s and early 1970s, fourteen more universities added faculties of science between 1965 and 1978, most of them offering at least a master's degree in addition to undergraduate instruction. Those without faculties of science mostly taught botany to lower-division undergraduates through their faculties of general education (*kyōyōgakubu*) or faculties of agriculture (*nōgakubu*).[8]

The first postwar decades were a time of restructuring and revival of prewar research programs at many of the former imperial universities, under conditions of sparse funding and usually inadequate equipment. The University of Tokyo created a faculty of general education at Komaba in 1949, subsuming the former No. 1 Higher School and the Tokyo Higher School, and provided instruction in plant sciences from the start. Of the many subfields of plant research, postwar Japanese scholars were especially active in the study of fungi, microbiology, and oceanography. At the main Hongō campus research on fungi recovered almost at once in 1945, leading to a mycology symposium in 1950 that turned into the Mycological Society of Japan (Nihon Kingakkai) in 1956. By 1977 this organization counted more than 1,000 members and published a high-quality house organ, *Mycoscience,* for readers of English. In 1952 the university added a research institute in applied microbiology, initially focused on taxonomy of the *Aspergillus* genus, comprising hundreds of mold species, and on practical applications for yeast. That same year the faculty of agriculture opened a research institute in applied microbiology. Research in the physiological chemistry of plants moved from the botany department to a new department of physical chemistry in 1958, later renamed cell physiology (*saibo seirigaku*). The university created an oceanography research institute in 1962 on the site of the former Tokyo Higher Normal School in Nakano, followed by research in marine botany starting in 1977.[9]

Initiatives such as these extended the University of Tokyo's edge as the premier locale for botanical research in the country, but other long-established campuses developed a number of specialties of their own. Kyoto University designated botany as a separate department within its science faculty in 1921, the same year it opened its Ashiu Forest Research Station (originally Ashiu University Forest). The university added a small botanical garden for research two years later and by 1929 featured three chairs in plant sciences. In the early years after World War Two the main research foci were plant physiology and ecology, plant cytology, and taxonomy. The university's agriculture faculty set up a timber research unit in 1944 and a foodstuffs research institute studying applied genetics in 1946, reflecting the needs of wartime and early postwar. By 1971 a plant reproduction research facility was operating within the agriculture faculty. A year later the faculty of science added a focus in environmental biology, drawing on new strengths at the botanical garden in botanical ecology.[10]

Tohoku University's curriculum in biology closely paralleled Kyoto's, with three chairs in botany as early as 1925. The main research facility, the Mount Hakkōda Botanical Laboratory (Hakkōdasan Shokubutsu Jikkenjo) in Aomori prefecture, opened in 1929 as an experimental station for Tohoku students to analyze forest plants and was used by other universities as well. In the early 1960s research at Tohoku University centered on plant taxonomy, physiology, and ecology.[11] In contrast, Kyushu University's agriculture faculty established its first chair in plant sciences in 1919 but abolished it in 1958 in favor of other areas of biology, leaving a lone position in plant physiology dating to 1950 in the university's college of science. As appointments in biology expanded from five to ten chairs in the 1960s, none resulted in adding more botanists; instead most instruction in that field was relegated to lower-division courses in the faculty of general education.[12]

Hokkaido University was founded in 1876 as Sapporo Agricultural College and became Hokkaido Imperial University in 1918. As early as 1904 the college established the 2,700-hectare Tomakomai Experimental Forest. During the interwar period Miyabe Kingo (1860–1951) and other scholars created a vibrant research cluster at the university in plant pathology, mycology, marine botany, and other specialties. Yamada Yoshio (1873–1958), an alumnus of Tokyo Imperial University who spent the years from 1928 to 1930 as a visiting scholar at the University of California, Berkeley, was the most prominent botanist at Hokkaido University in the early postwar period. Yamada wrote more than 100 papers on seaweeds and headed a research center on marine plants at the university, the largest in Japan. In 1952 he became the founding president of the Japanese Society of Phycology (Nihon Sōrui Gakkai, for the study of algae), which from then to the present has published the quarterly journal *Sōrui—Japanese Journal of Phycology*.[13] Hiroshima University, established as Hiroshima Bunrika University in 1929; Osaka University, designated an imperial university in 1931 based on Osaka Prefectural Medical School (1869); and Nagoya University, founded in 1939 as Nagoya Imperial University, all developed broad biology programs with some level of instruction in plant sciences in the early postwar period. Hiroshima University opened its Miyajima Natural Botanical Garden (Miyajima Shizen Shokubutsu Jikkensho) in 1964 on the northern edge of Miyajima (Itsukushima) island, with a focus on subtropical plants, marine vegetation, and mosses. Nagoya University formed a faculty of agriculture in 1951 with four professors expert in botany; by 1974 its science faculty included three chairs in this field.

The University of Tsukuba, northeast of Tokyo in Ibaraki prefecture, traces its origins to four academies in science, education, agriculture, and physical education that merged in 1949 to form Tokyo University of Education. This university, in turn, served as the foundation for the University of Tsukuba (1973), which quickly became the centerpiece for more than 300 schools and research institutes based in Tsukuba Science City. As a new national university, the University of Tsukuba focuses on science, technology, engineering, mathematics, and medicine. The university incorporated the three chairs in botany from the former Tokyo University of Education and in 1977 took full control of the Sugadaira Montane Research Center (Tsukuba Daigaku Sugadaira Kōgen Jikken Sentā) in Nagano prefecture, a colonial-era institute founded in 1934 to conduct agrobiological research in weather conditions similar to those in Manchukuo. This center is the largest of Japan's four research facilities specializing in mountainous areas.[14] The Tsukuba Botanical Garden, a unit of the National Museum of Nature and Science in Tokyo, opened in 1983 and provides a major research locale for plant specialists at the University of Tsukuba. Only the University of Tokyo, with its sizable herbarium as well as the Koishikawa Botanical Gardens, offers more

Figure 6.1 National Museum of Nature and Science Tsukuba Botanical Garden, Ibaraki prefecture. Savanna house (*left*), tropical greenhouse (*right*), 2017.

extensive infrastructure for botanical research in Japan. Private universities such as Waseda, Keio, and Doshisha, which receive less public funding for botanical research than do national universities, offer instruction in various aspects of plant sciences, often through programs in ecology and environmental studies, but mainly focus on biomedical sciences.

Dozens of university specialists resumed their investigations of overseas research sites as soon as financial and travel restrictions abated following the American occupation of Japan, between 1945 and 1952. Research teams prepared nearly two dozen descriptive inventories of vegetation in Micronesia, Nepal, the Himalaya, Brazil, the Ogasawara islands, Southeast Asia, and various tropical rain forests. Hara Hiroshi of Hokkaido University, Kanai Hiroo of the National Science Museum, and Ōhashi Hiroyoshi of Tohoku University assembled herbarium specimens from India, where Hara surveyed plants starting in 1960. Iwatsuki Kunio, who moved from Kyoto University to the University of Tokyo in 1982, continued his inventories of plant diversity in Southeast Asia, while Tatsumi Tatsuo led scientific surveys of the Svalbard peninsula, Norway, between 1983 and 1988.[15] These and other forays abroad set the stage for younger Japanese scholars to investigate plant life internationally in the early twenty-first century.

Scholarly societies are the backbone of intellectual life and the locus for exchanging research ideas in plant biology as in other scholarly disciplines in Japan. The Botanical Society of Japan (Nihon Shokubutsu Gakkai), founded in 1882 as the Botanical Society of Tokyo, assumed its current name in 1931 and in October 1957 celebrated its 75th anniversary with a convention of 1,151 members held at the University of Tokyo. By the 100th anniversary in 1982 the society had moved its offices to their current quarters in the Esperanza (formerly Hongō Tōshin) building in Hongō adjacent to the University of Tokyo campus. By far its most important activity is to publish the bilingual *Journal of Plant Research* (*Shokubutsugaku zasshi*), which originated in the society's earliest days. For many years this publication was the premier outlet for research by Japanese scholars, but in the past twenty years authors from elsewhere in Asia and throughout the world have been increasingly represented in its pages. A number of more specialized botanical associations were founded after World War Two, including the Japanese Society of Plant Physiologists (Nihon Shokubutsu Seiri Gakkai), established in 1959 to bring together specialists in botany, ecology, and agronomy, and the Japanese Association for Photosynthesis Research (Kōgōsei Gakkai), founded in 1980 with 300 members. The Photobiology Association of Japan (Nihon Kōseibutsugaku Kyōkai) began two years later as an umbrella group of fourteen academic societies working in photobiology, including botany and photosynthesis.[16] Several more specialized societies have been formed since.

The Japan Association of Botanical Gardens was founded in 1947 and today counts about 150 member gardens and arboretums, of which 50 are considered major facilities for plant research.[17] Other members include alpine reserves, tropical gardens, city parks, flower parks, public gardens such as Shinjuku Gyoen and the Kobe Municipal Arboretum, and numerous sites for people's recreation, some of which also contribute to botanical scholarship.[18] Gardens maintained by pharmaceutical companies and other businesses conduct specialized research, especially on medicinal plants. The Japan Association of Botanical Gardens meets annually and sponsors research workshops and training institutes quarterly, as well as study tours of botanical sites abroad and an exhibit each year at the gigantic Tōbu department store in Tokyo's Ikebukuro district.[19] Despite the presence of these various sites, the leading plant specialist and science administrator Iwatsuki Kunio believes that as research facilities, both Japan's science museums and its botanical gardens have "a very frail foundation."[20]

As the nation's premier space for botanical research, the sixteen-hectare Koishikawa Botanical Gardens and ten-hectare branch at Nikkō suffered comparatively slack times from the end of World War Two until 1978, when a spate of activity in seed and cell preservation was accompanied by plant inventories, management restructuring, and planning for future research projects. As of 1983 the garden's annual operating budget was ¥20 million ($180,000 U.S.), a figure that remained static for years and proved quite inadequate to maintain the grounds for research. By contrast, at the same time the Jindai Botanical Park in the Tokyo suburbs had a budget of ¥500 million ($4.5 million) to operate its forty-eight-hectare space, while the Hiroshima Botanical Garden, at 18.3 hectares, was budgeted at ¥300 million ($2.7 million). In the 1980s the Koishikawa facility increased its research and instructional staff to six specialists and introduced technology for molecular analysis. New research on endangered plant species and preserving biodiversity took place, increasingly by engaging with scholars in science and agriculture faculties at the Hongō campus. By the early 2000s the instructional and research staff had doubled to twelve specialists, whose responsibilities included supervising graduate students from Japan as well as from East and Southeast Asia.[21] Herbarium specimens are divided between the botanical garden and the university museum at Hongō; they comprise Japan's most extensive collection, totaling about 170 million items, primarily from East Asia and the Himalaya.[22]

Also in central Tokyo is the Yumenoshima Tropical Greenhouse Dome (Yumenoshima Nettai Shokubutsukan), a soaring 28-meter triple-dome structure containing 1,000 species of aquatic and land plants, including 35 species found naturally only on Japan's semitropical Ogasawara islands. Built in 1988 on a landfill in Tokyo Bay, the facility aims to preserve the fifteen Ogasawara

Figure 6.2 Triple hemispheres of the Yumenoshima Tropical Greenhouse Dome, the main attraction on Yumenoshima (island of dreams), a landfill in Tokyo bay with numerous athletic facilities and a museum, 2017.

Figure 6.3 Spectacular interior of the Yumenoshima Tropical Greenhouse Dome, Tokyo, 2017.

species considered endangered. Displays are heated by circulating hot water fired by a nearby garbage incinerator and kept at twenty-four degrees centigrade, recreating a tropical rain forest with giant water lilies, palms, vines, orchids, ferns, mangos, and star fruit. The Yumenoshima Dome combines nature education for the general public with research facilities for Tokyo-area specialists interested in tropical plants.[23]

Elsewhere in the Tokyo region are a number of nature gardens for specialized research in plant biology. The 1-hectare Akatsuka Botanical Garden (Akatsuka Shokubutsuen) in Itabashiku, founded in 1981, contains more than 600 varieties of trees and plants, including medicinal herbs mentioned in Japan's oldest poetry collection, the *Man'yōshū* (Collection of ten thousand leaves, 759). This hilly, well-organized garden features metasequoias grown from seeds of trees discovered in western China in 1946.[24]

Figure 6.4 Even in late autumn the Akatsuka Botanical Garden, Tokyo, displays vibrant flowers and foliage, 2017.

Figure 6.5 Aritaki Arboretum, Koshigaya, Saitama prefecture, contains a trove of Asian tree species, some of which are seldom seen elsewhere, 2017.

The Aritaki Arboretum (Aritaki Āborētamu) on a long, thin "spaghetti lot" in Koshigaya, Saitama prefecture, contains wide-ranging specimens of trees collected throughout Asia by its founder and curator, Aritaki Tadahiko.

The Institute for Nature Study (Shizen Kyōikuen) at Shirokanedai was created in 1949 on twenty hectares originally belonging to Matsudaira Yorishige (Sanuki no Kami, 1622–1695), daimyo of the Takamatsu domain. The site became an imperial estate in 1917, administered by the National Museum of Science. In 1962 the museum absorbed the Institute for Nature Study, with its 500-year-old pines and chinquapins, more than 800 plant species, and special gardens for vegetation from the Musashi plain and roadside plants. Its public lectures on ecology and strong emphasis on outreach accompany the investigations of scientific researchers.[25]

Figure 6.6 A branching Chinese wingnut (*Pterocarya stenoptera*) in the Aritaki Arboretum, Koshigaya, Saitama prefecture. This towering species is widely found in East Asian landscape gardens, 2017.

The Jindai Botanical Park (Jindai Shokubutsu Kōen) in suburban Chōfu was established in 1961 on the site of a wartime youth drill ground and vegetable garden that after 1945 became a seedling nursery for Tokyo street trees. The park currently counts more than 100,000 plants representing 4,800 varieties, including aquatic plants. Although mainly a 42.5-hectare horticultural space featuring elaborate displays for public enjoyment, the park focuses its research on protecting endangered species.[26] A half hour west of Jindai is the Tama Forest Science Garden (Tama Shinrin Kagakuen) on fifty-seven hectares of woodland at the foot of Mount Takao in Hachiōji, Tokyo. Once a noted battlefield during the feudal wars of the mid-sixteenth century, this hillside venue became an imperial forestry experiment station in 1921 and today is managed by the government's Forestry and Forest Products Research Institute.

Figure 6.7 Jindai Botanical Park, opened in 1961 in western Tokyo, contains a vast barrier-free rose garden, 2009.

The garden contains 6,000 trees of about 600 domestic and foreign species, including 2,000 cherry trees representing 250 varieties. This facility, which focuses on molecular systematics, is a premier locale for dendrological research and experimentation in groves of older trees that have been in existence for 150 years or longer.[27]

Nearby is the Tokyo University of Agriculture Botanical Garden (Tokyo Nōgyō Daigaku Shokubutsuen) on a broad hillside within view of Mount Fuji in Atsugi, Kanagawa prefecture. Here the focus is on economic and medicinal plants from around the world, including 310 species of wild plants.[28] A similar but less inviting research facility is the Kemigawa Arboretum (Ryokuchi Shokubutsu Jikkensho, lit. Experimental Station for Landscape Plants), established in 1965 by the University of Tokyo agriculture faculty in Hatamachi, Chiba prefecture. This small garden contains about 350 species of woody plants and flowers and is operated as a research and breeding facility in a somewhat inaccessible location rarely open to visitors.

The elaborate Tsukuba Botanical Garden (Tsukuba Jikken Shokubutsuen), operated since 1983 by the National Museum of Nature and Science, covers 14 hectares in Ibaraki prefecture and conserves more than 7,000 plant species ranging from alpine to tropical vegetation. Its four greenhouses include savanna plants, tropical rain forest species, aquatic plants, and economically useful plants. Its forest areas contain broadleaf evergreens, conifers, and decid-

Figure 6.8 Extensive hillside fields and greenhouses of the Tokyo University of Agriculture Botanical Garden overlooking the city of Atsugi, Kanagawa prefecture, 2017.

uous broadleaf trees representing various climate zones within Japan, as well as a promenade lined with metasequoias. A special focus is endangered species and biodiversity. Research topics center on molecular DNA sequencing and morphological, cytological, and chemical data. Although in Japan "most scientists have a weak consciousness of public outreach,"[29] the Tsukuba garden operates a number of public programs for botanical education. Although open to all, attendance is low, partly because the facility is removed from Tsukuba station, which in turn is a forty-five-minute trip by express train from Akihabara in central Tokyo. The National Museum of Nature and Science has moved its entire plant research department to Tsukuba, making the garden a highly productive center of scholarship in plant biology, in association with specialists from nearby University of Tsukuba.[30]

In addition to the Hokkaido University Botanic Garden (est. 1886), discussed in chapter 4, a major research facility in northern Japan is the Tohoku University Botanical Garden, established in 1958 on Mount Aoba surrounding the site of a castle built in 1602 by the daimyo Date Masamune (1567–1636), a storied warrior and founder of the city of Sendai. Mount Aoba was used as an army base from 1868 to 1945, although in the 1930s botany students from Tohoku University could enter for fieldwork. Because its forested areas were scarcely disturbed by humans for more than three centuries, the forty-nine-hectare garden is home to flora and fauna of great academic value, such as 700 species of vascular plants, 140 species of mosses, and some firs larger than 150 centimeters in diameter and taller than 40 meters. Also included are 1,300 willows representing 200 species and an herbarium. The garden is a major sightseeing attraction in Sendai, and as the most important botanical collection between Tsukuba and Hokkaido it draws researchers studying the regional vegetation of northeastern Japan.[31]

Long overshadowed by its adjacent zoo, the Higashiyama Botanical Park (Higashiyama Dōshokubutsuen) in Nagoya, established in 1937, contains more than 7,000 varieties of plants and Japan's oldest greenhouse. Flowering species are prominent, including bougainvillea, hibiscus, tropicals, and insect-devouring plants. Japanese and Asian vegetation is featured on the park's extensive grounds. Although Higashiyama is primarily a strolling park for recreational enjoyment, the huge variety of angiosperms includes species seldom found elsewhere and attracts researchers from throughout the country.[32] Also popular with visitors, although especially oriented toward biological research, is the Kyoto Botanical Garden (Kyoto Furitsu Shokubutsuen) along the Kamo river in the heart of Kyoto. Established in 1924 on a twenty-four-hectare site, the garden was commandeered by occupying United States military forces right after World War Two and used for housing for more than ten years, during which most of the plants were cut down. The Americans returned the property in 1957, and the garden reopened four years later after extensive repairs. Its goals include showcasing the ecology of various districts of Japan, preserving the surviving trees in the garden, and displaying plants from abroad in its huge greenhouse to underscore biodiversity. The garden counts more than 120,000 plants in 12,000 species, including 4,500 species in the conservatory.[33]

Osaka City University established its botanical garden (Osaka Shiritsu Daigaku Rigakubu Fuzoku Shokubutsuen) in 1950 on 26 hectares in suburban Osaka as a research facility with special focus on 450 species of native plants and a large arboretum area representing 11 types of Japanese forest ecosystems. The garden also contains the first metasequoias grown from seeds introduced to Japan from China after World War Two. Systematics, genetics, dendrology, and physiology are key research areas, with particular attention to preserving endangered plants from the Osaka region.[34] The Kobe City Arboretum (Kobe Shiritsu Shinrin Shokubutsuen, also called Kobe City Forest Botanical Garden) was established in 1940 and opened in 1964 on a huge 143-hectare hillside property along the Nishi Rokkō Driveway in northern Kobe. Today it features about 1,200 species of trees and shrubs from around the world and attracts dendrological researchers in all seasons.[35] Further west is the Hiroshima City Botanical Garden (Hiroshimashi Shokubutsuen), opened in 1976 on 18 hectares overlooking the Seto inland sea. The garden has facilities for research and conservation and counts nearly 250,000 individuals representing 11,700 species in its collection, including about 1,200 species of tropical and subtropical plants in its six greenhouses.[36] These and other botanical gardens serve the public enthusiasm for horticulture while also acting as field laboratories for university researchers in plant biology. In this way, university scientists, the scholarly societies they populate, botanical gardens, and government financial assistance form the main infrastructure for botanical research in Japan.

Arima Akito, president of the University of Tokyo, pointed out in 1992 that government support for basic research stagnated in the 1980s, with particular impact on ten leading national universities that produced 47 percent of the 92,363 physics papers published in Japan between 1976 and 1986.[37] In Arima's view, approximately thirty Japanese universities were "strong in scientific research as well as education, but they are suffering from lack of adequate funds" for research. Primary through baccalaureate education were relatively well funded, he added, but "to provide incentives and financial support, universities must create more scholarships for graduate students."[38] Writing also in 1992, Hirano Yukihiro of the National Institute of Science and Technology Policy argued that "Japan remains conspicuous for its low level of government funding of R&D." Japanese businesses increased their basic research activity after 1980, but "private sector research inevitably involves commercial restraints which can, on occasion, compromise creativity." Hirano called for greater public and private support in almost all areas of basic research, noting that Japanese research was particularly "weak in the life sciences and oceanographic-earth sciences," to the point where "many excellent Japanese scientists are choosing to take their research abroad,"[39] a practice that continues to this day.

As in many other mature economies globally, public financial support for higher education in general and scientific research in particular eased in the face of other state budgetary priorities early in the twenty-first century. Overall university funding from Japanese government sources fell about 1 percent per year between 2004 and 2017, while the premier scientific research institute Riken (Rikagaku Kenkyūjo) saw its budget decline by 20 percent between 2007 and 2017. Partly for want of financial support and a lack of jobs when they finish, the number of doctoral students in all fields shrank by 18 percent between 2003 and 2018. Igami Masatsura of the National Institute of Science and Technology Policy said in May 2017, "Japan is not challenging itself to take on new scientific areas, and it seems that the research field is becoming rigid."[40] One reason was that the state's overall science and technology budget declined about 5 percent between 2012 and 2017, prompting concern about the global standing of Japanese universities as Japan's share of highly cited papers stagnated and the total number of papers published by scholars at universities and research institutes grew by just 14 percent between 2005 and 2015, while China's output doubled at the same time. A May 2019 study revealed that China ranks second to the United States in the quality of its scientific research, whereas Japan's standing has fallen over the past two decades.[41] Basic research, including plant biology, was especially hard hit by the budget cuts. Although Japan had the second most Nobel science prizewinners after the United States during the early twenty-first century, insufficient funding almost assures that the country will have great difficulty holding its own in scientific accomplishment and international recognition in the future.[42]

Prompted by major increases in research spending by both China and South Korea, Japan's Council for Science, Technology, and Innovation announced in January 2018 a 7 percent increase in government spending on science and technology, for a total outlay of ¥3.84 trillion ($35 billion). Some of the budgetary increase was attributable to changes in how expenditures were calculated; much of the rest was earmarked for artificial intelligence and other areas of technological investment rather than for basic science. The Abe Shinzō government pledged to lift public spending on science and technology to 1 percent of gross domestic product by 2020, yet allocations of operating funds for national universities remained unchanged for 2018 at just over ¥1 trillion—a pyrrhic victory after 15 years of 1 percent cuts per annum.[43] Apart from support from drug manufacturers for studies of medicinal plants, little private funding is available to botanical researchers despite efforts by universities to build private, corporate, or overseas partnerships to support basic research.

Further Diversity and Specialization

Scientific research in Japan during wartime and the early postwar years underwent the same privations as society as a whole. Between 1937 and 1955 the national economy experienced eighteen years of no net growth, even though the population increased by 25.3 percent at the same time. As late as the mid-1960s many universities, research institutes, and botanical gardens suffered from inadequate financial support, shopworn facilities, and a lack of direction bordering on inertia carried over from wartime. In plant biology, postwar research on morphology—a staple of prewar training—increased very slowly compared with countries in Europe and North America, although Maekawa Fumio (1908–1984), a specialist on leaves and seasonal flowers, trained a number of specialists at the University of Tokyo.[44] Other centers of morphological scholarship included Tohoku, Kyoto, Fukui, Chiba, and Hirosaki Universities. By the 1980s lectures in anatomy and morphology became so specialized that the field of plant morphology grew increasingly attenuated.[45]

Japanese plant physiology, an equally time-honored area of inquiry, likewise underwent specialization after World War Two; unlike morphology, it grew robust and reached an "international level" of achievement in photosynthesis, plant growth and movement, cell physiology, and regenerative physiology.[46] Hokkaido University became a vigorous site of physiological scholarship, led by Sakamura Tetsu (1888–1980) and his students working on cold-resistant plants. At Osaka University, Kamiya Noburō (1913–1999), who studied at the Free University of Berlin after the war, led research on protoplasm mobility and electric physiology.[47] Similar training in physiology grew vigorous in the 1980s at the University of Tokyo, Hiroshima University, University of Tsukuba, and Tokyo Metropolitan University. Research on plant hormones took place at the University of Tokyo, Tohoku University, Kyoto University, University of Tsukuba, International Christian University, and elsewhere, so that Japan rose to the forefront internationally in certain areas of plant physiology.[48]

Research in molecular biology after World War Two turned increasingly to biochemical genetics, studying cell nuclei and the mechanisms of sudden mutations. The National Institute of Genetics (Kokuritsu Idengaku Kenkyūjo) was established in 1949 and helped introduce plant genetics four years later. Especially noted was the work of molecular biologist Kutsukake Kazuhiro at Okayama University on the regulatory mechanisms of gene expression.[49] Kutsukake is particularly well known for identifying reversal enzymes in genetic research. The University of Tokyo, Osaka University, and Kyoto University were other pioneering institutions in this new field of microbiology.[50] The genetics institute added a department of molecular biology in 1969, recognizing that this field—first developed in the 1930s—had entered the mainstream of biological research by the 1960s, including plant biology.[51]

When the American molecular biologist Herbert Boyer of the University of California, San Francisco, and geneticist Stanley Cohen of Stanford University announced the development of recombinant DNA (deoxyribonucleic acid) technology in June 1973, scientists and governments around the world hastened to draft research guidelines based on self-regulation to assure safety management against biohazards in basic research as well as in industrial applications.[52] In Japan, molecular biologist Watanabe Itaru of Keio University and geneticist Tajima Yatarō of the National Institute of Genetics led a group of scientists in cooperating with the Ministry of Education and the Science Council of Japan to develop appropriate policies. Life science specialists were understandably eager to introduce recombinant DNA technology but initially faced heartfelt opposition from scholars in the humanities and social scientists, who raised ethical questions. Finally in March 1979 the ministry announced guidelines that proved to be stricter than those adopted in December 1978 by the National Institutes of Health in the United States. The Institute of Medical Science at the University of Tokyo opened a Genetic Analysis Center in 1980, followed by new facilities at 67 other universities and 7 research centers to implement the Japanese guidelines, making "possible huge advances in recombinant DNA research during the 1980s."[53] Among the most significant centers was the National Institute of Genetics in Mishima, Shizuoka prefecture, which was reorganized as an interuniversity facility for joint use by universities in 1984. That same year the institute created a DNA Research Center with four laboratories, followed by a DNA databank of Japan in 1987.[54]

Sometimes criticized as unscientific, the venerable field of taxonomy in Japan was increasingly overshadowed by new developments in plant studies such as molecular biology and DNA analysis. Writing in the 1980s, Emperor Hirohito noted that taxonomy had been based on morphology, but "now we can classify plants at the molecular level based on DNA. New species are likely to emerge via molecular biology,"[55] an accurate prediction. Much emphasis was placed on biodiversity during the mid-1980s, especially at the Koishikawa Botanical Gardens, where researchers flocked to learn new developments in plant classification. Once genetic information became usable, molecular systematics became mainstream, and deployment of the Angiosperm Phylogenetic System (APG III) grew widespread. Information on DNA groupings helped clarify plant lineages and overturned many previous classifications, to the point that today the debates over angiosperm taxonomy are largely settled, as is true in morphology, physiology, and other botanical specialties. Botanical gardens have switched their nomenclature, and their signage, to reflect the new DNA-based classifications but generally also show the old names in parentheses.[56] Examples of scholars relying on DNA research include Setoguchi Hiroaki of Kyoto University, who studied seven species of the genus *Aucuba* (spotted laurel) to show that the oldest distribution is in south China, and Fujii Noriyuki of Tokyo Metropolitan University, who tested forty-two taxa of Japanese alpine plants in an effort to determine whether they moved south in the glacial age.[57] Chromosome research, long a part of taxonomy, was similarly transformed by new botanical techniques analyzing chemical components and their relationship to taxa, including studies of isozymes. The researcher Nishikawa Tsunehiko counted the chromosome numbers of 477 taxa in 56 families, 232 genera, and 414 species of flowering plants in Hokkaido. The electron microscope is a powerful tool for magnification in systematics; computers assemble data such as those gathered by Nishikawa for identifications and classifications.[58]

As in other countries, Japan experienced considerable public opposition to genetically modified crops, regardless of the 1979 safeguards on recombinant DNA technologies and other government restrictions, including safety evaluations and environmental impact assessments. Some critics said, "Gene modification technology, like nuclear weapons, is humankind's greatest enemy."[59] Japan has been slower than many countries to develop genetically modified foods for public sale, even though Japanese eat genetically modified products made from soybeans and corn imported from abroad.[60] Satō Yōichirō of Kyoto Prefectural University has used isozyme and DNA analyses since the 1980s to analyze the gathering of wild rice and later cultivation of rice in prehistoric India, Southeast Asia, and China, finding that Indica and Japonica varietals belong to "two different races."[61] At the Riken research center in 2012, head of the radiation biology laboratory Abe Tomoko and her team used a particle accelerator for crop mutations, seeking to modify rice genes for resistance to salt. Two years later her laboratory became the world's first to use heavy-ion accelerator technology to modify wheat genes for plant breeding. The time span for breeding varieties of crops, trees, and angiosperms using heavy-ion beams, she reported, was two to three years, much faster than through hybridization or gene recombination.[62] By beaming carbon ions at branches of the Keiōzakura 13 strain, her team created a new variety of cherry tree that can bloom year-round without requiring the usual prolonged bouts of cold weather to blossom correctly.[63]

Other Japanese botanical researchers conducted experiments using new technologies to produce unprecedented results. Wakita Kōichi, commander of the International Space Station, returned in 2009 with a cherry pit from a 1,250-year-old cherry tree in Gifu prefecture sent into space the previous year. Plant biologist Yoshimura Takao grew a four-meter tree from the pit within four years, not the ten years normally required from seed to blossom, presumably because of exposure to high-energy radiation in space; the press dubbed this precocious tree the rapid-growth "space cherry."[64] Scientists at the National Institute for Basic Biology (Kisō Seibutsugaku Kenkyūjo) in Okazaki, Aichi prefecture, revealed in 2014 that they had brought back to life a long-lost yellow morning glory by implanting a snapdragon gene in a creamy-flowered morning glory.[65] The following year biochemists at Nagoya University developed a reagent called ClearSee from twenty-four chemical compounds to make plants translucent by soaking them in the reagent for four days, which removes the chlorophyll and allows cells to be seen under a microscope without cutting them open.[66] Yonekura Kōji at Riken and biologists elsewhere now use cryo-electron microscopy (cryo-EM) to create three-dimensional images of large and complex protein molecules and related applications, an innovation that is expected to transform biological research in the twenty-first century.[67]

Waves of diversity and specialization washed over other more established areas of Japanese botanical research after 1945 as well. Thanks in part to research on Cenozoic (Tertiary, 66–2.8 million years ago) fossil flora via pollen analysis, early postwar dendrologists established a nearly complete tabulation of Japanese gymnosperms. Studies of plant chromosomes and similarities in biochemical composition allowed more precise classification of tree lineages, while new ethnobotanical studies brought to light local names of plants and shrubs. Scholars examined ecosystems via statistical methods that classified forest communities, yet Japan counted very few forest physiologists, so older approaches to silviculture carried over from prewar times. Much attention was paid to forest fungi and blights, resulting in effective control measures and less loss of nursery stock once fifty species of wood-rotting fungi were identified.[68]

After World War Two a large number of studies of mushrooms and other fungi appeared in Japan, including papers on mating and breeding. For example, until 1978, sixty-five taxa of the genus *Cordyceps* (a fungus parasitic on insects that has certain medicinal benefits) were known in Japan; that year Kobayashi Yoshio (1907–1993) and Shimizu Daisuke (1915–1998) reported 6 species new to science, and 9 taxa new to Japan.[69] More recently 2 species of Plectomycetes (a class of fungi releasing spores during decay or disintegration) and 59 species from 15 families of Discomycetes (a now discontinued class of fungi), 8 of which were new to Japan, were collected by Nagao Hideyuki of the Universiti Sains Malaysia and Fukiharu Toshimitsu of the Natural History Museum and Institute, Chiba.[70]

Studies of mosses and lichens, in virtual abeyance during World War Two, revived thanks in part to the Hattori Botanical Laboratory, founded in 1946 at Nichinan, Miyazaki prefecture, by liverwort specialist Hattori Shinsuke (1915–1992). This facility initially focused on Japanese moss taxonomy and genetics, and then in 1963 repositioned itself as an institute for studying mosses and lichens around the Pacific rim. It publishes two journals and other occasional works and includes scholars from abroad among its researchers. Specialists from the Hattori laboratory, Hiroshima University, and elsewhere formed the nucleus of the Bryological Society of Japan (est. 1972), which publishes a journal, newsletter, and other studies for worldwide circulation.[71] Mosses are popular topics for public botany in Japan, where the Northern Yatsugatake Moss Association in Nagano prefecture has organized moss hikes since 2011 for amateur botanists. In 2018 the country's most prominent bryologist, Higuchi Masanobu of the National Museum of Nature and Science, led monthly tours of Yatsugatake from May to October for 140 enthusiasts at a time.[72] Mosses are highly popular for botanical ornamentation: Near Nikkō, Kiyomura Oichi, Japan's "Moss King," harvests mosses from large tracts he owns for sale to nurseries, temples, and rooftop gardens.[73]

Marine botany is one of the most well-established fields of scholarship in Japan, tracing its origins to Yatabe Ryōkichi at Tokyo Imperial University and his pupil Okamura Kintarō (1867–1935), the "Father of Marine Botany in Japan." Okamura worked for the Imperial Institute of Fisheries (Suisan Kōshūjo, now Tokyo University of Marine Science and Technology), identifying about 1,000 species of seaweeds before he retired in 1931.[74] From Hokkaido University to the University of Tokyo, Kagoshima University, and Okinawa Institute of Science and Technology, marine botany and algae studies thrived in the postwar period. Not long after the war, paleogeologist Konishi Kenji identified a new genus, *Anatolipora* (*Anatolipora carbonica konishi*), in fragments of algae found in limestone from the lower Carboniferous period (359–323 million years ago) in Niigata prefecture.[75] A major milestone was the 1977 publication of *Nihon tansuisō zukan* (Illustrated Japanese freshwater algae), a 933-page compendium edited by Hirose Hiroyuki and Yamagishi Takaaki that glossed 2,308 species and variants. This work complemented Okamura Kintarō's prewar publications on marine algae. By the end of the twentieth century Japanese scholars had identified more than 5,000 marine and freshwater plants.[76]

Plant biologists working on herbal remedies in the postwar period have found employment with pharmaceutical firms that produce the 165 ingredients currently approved to be used in herbal medicines in Japan. As of 1961 the herb garden of the Takeda Pharmaceutical Company, the country's largest drug manufacturer, featured the cultivation of 29 species of the *Digitalis* genus for treating heart conditions, and 120 species of the *Allium* genus for the study of thiamine derivatives. By then the company exchanged seeds with botanical gardens and research institutes in thirty-eight countries, a sign of the internationalization of herbal remedies that has grown greatly since.[77] Today Tsumura and Company is the leading Japanese manufacturer of herbal medicines, producing 128 of the 148 remedies approved for reimbursement under the national health insurance scheme. Chinese licorice root (*Radix glycyrrhizae*) and ginger (*Zhingiberis rhizome*) are the most commonly used herbs. Arai Ichirō of Tsumura pointed out in 2009 that Japanese manufacturers relied on imports for 97 percent of the herbal materials used for medicine and food, mainly from China, where demand for Japanese herbal remedies has soared in recent years.[78] Nippon Shinyaku Company, another large pharmaceutical firm, operates the Yamashina Botanical Research Institute (est. 1934) in Kyoto, which as of 1963 counted 1,200 species of 600 genera in its collection of economic plants, both spices and components of manufactured drugs. The company collects plants throughout Asia and conducts exchanges with botanical gardens internationally. Today the Yamashina institute contains more than 3,000 plant species.[79] Also serving both plant biologists and farmers is the Usui Seed Bank for economic plants operated in Ikeda, Nagano prefecture, by Usui Kenji and Usui Tomoko.[80]

Ōba Hideaki and David E. Boufford pointed out in 2005 that biogeographical relationships and the evolutionary relationships of angiosperms "are among the hottest topics in biology."[81] By then the early postwar outpouring of encyclopedic reference works in plant biology had built a sturdy foundation for the specialized studies that characterized much Japanese botanical research in the ensuing decades.[82] Regional distribution maps became common research products, from Hara Hiroshi's maps of Japanese seed plants (1949, 1977, 1979) to Kurata Satoru's forestry maps (1971–1976), Horikawa Yoshio's field site surveys of Japanese flora showing the distribution of 200 species (1972–1976),[83] and the many regional studies since then that crowd the biology shelves of major Japanese bookstores. Examples of regional research included an account of primrose reproduction among 2,000 plants in Ajigaura, Chiba prefecture, between 1978 and 1983, and a survey of flora and fauna in Fujioka, Gunma prefecture, from 1983 to 1988 that identified 143 plant genera and 1,223 species.[84]

Specialized topical research proliferated after 1945, such as Itō Kōji's study of vegetation in salt marshes in northern Hokkaido, completed in 1963, and Kurata Satoru and Nakaike Tsuyoshi edited an eight-volume illustrated history of ferns in 1979–1997, *Nihon no shida shokubutsu zukan* (Illustrated history of Japanese ferns), on behalf of the Japan Pteridological Society (Nihon Shida no Kai, est. 1957).[85] Hara Hiroshi produced detailed studies of the honeysuckle family (*Caprifoliaceae*), including the genus *Viburnum,* with distribution maps. He noted that researchers now examined this family not only via morphology but also cytogenetically and chemotaxonomically.[86] Okita Yoshihiro focused on dwarf palms, which came to Japan in the early Edo period from Southeast Asia and are easier to grow than bonsai plants. Half these potted evergreens are striped and half solid green, flourishing either indoors or outside, whereas bonsai mostly need to live outdoors.[87] The leading female botanist Kuginuki Fuji in 2000 developed theories for conducting statistical analyses of lower epidermal tissues on 727 rhododendron subgenera, in order to advance the classifications of species.[88] Paleobotanists too have engaged in specialized studies since the 1960s, focusing especially on northeast Honshu sediments rich with plant megafossils dating to the late Miocene epoch (11–5.3 million years ago). The main finds have been deciduous broadleaf trees led by beeches, birches, and oaks, mixed with evergreens and a number of angiosperms.[89]

Plant Ecology in Postwar Japan

"People could not exist on this earth without plants. While living together with plants on the planet, here we'd like to consider the sorts of scholarly approaches that are being taken."[90] With these words the 134-year-old Botanical Society of Japan introduced its authoritative *Encyclopedia of Botany* (*Shokubutugaku no hyakka jiten*) in 2016, signifying the prime importance now accorded to ecological perspectives in botanical research. As discussed in chapter 5, Japanese study of ecology made a slow start in the early twentieth century despite the pioneering efforts of Miyoshi Manabu and his students, accompanied by the idiosyncratic ecological ideas of Minakata Kumagusu. The Association of Democratic Scientists (Minshushugi Kagagusha Kyōkai, or Minka), established in 1946 with sponsorship by the newly legalized Japan Communist Party, held yearly symposia on ecology from 1949 to 1957 as a part of its focus on the peaceful uses of science. In 1965 Minka became the basis for the larger Japan Scientists' Association (Nihon Kagakusha Kaigi), which today continues to emphasize the social responsibility of natural and social scientists through its chapters in each of Japan's forty-seven prefectures.[91] The establishmentarian Botanical Society of Japan held various symposia in the 1950s but included few panels on ecology. Later it held joint events with the Ecological Society of Japan (Nihon Seitai Gakkai), which was established in 1953 with 600 members. The Ecological Society began publishing the *Japanese Journal of Ecology* (*Nihon Seitai Gakkaishi*) in 1954, absorbing the fledgling *Plant Ecology Society Report* (*Shokubutsu Seitai Gakkaihō*) produced at Sendai since 1951.[92] Today the Ecological Society publishes two other journals as well, as do a dozen or more university research centers and specialized associations of plant scientists. Best known is probably *Hikobia*, produced since 1950 by the Hiroshima Botanical Club at Hiroshima University (Hiroshima Shokubutsugaku Kenkyūkai).

The University of Tokyo and Tohoku University were the leading academic centers of ecological research in plant biology before World War Two, the latter in good part because of the Mount Hakkōda Botanical Laboratory and the botanical garden on Mount Aoba in Sendai. In 1954 Monji Masami (1914–1977) succeeded Miyoshi Manabu (1862–1939) and Nakano Harufusa (1883–1973) as the beacon of botanical ecology at the University of Tokyo.[93] Early ecological research in Japan was often quite theoretical and highlighted plant communities created by humans, including Nakano's work on physiological ecology, whereas Monji shifted focus to fieldwork on the role of light in the production of plant material in natural communities under the severe environmental conditions of the 1970s. Monji retired in 1975 and was succeeded by Saeki Toshirō (1927–2004), winner of the Ecological Society of Japan's Asahi prize in 1978 for his work on plant growth via photosynthesis, based partly on research conducted in Australia between 1962 and 1964.[94] From Saeki's time onward ecology became an integral part of Japanese research in environmental science.[95] At Tohoku University, Yoshioka Kuniji (1910–1977) led an emergency survey of Japanese flora commissioned by the Cultural Agency of the Education Ministry in 1967, which resulted in the first national vegetation map. Yoshioka's work is summarized in his posthumous *Shokubutsu seitai ronshū* (Views of plant ecology, 1978).[96]

Other universities cemented their commitments to ecology, including plant biology, from the 1950s through the 1970s, priorities that have only solidified since. Both Osaka City University and Kyushu University began teaching plant ecology in 1950, followed by Tokyo Metropolitan University in 1956. Ecology was taught mainly at institutions with masters degree programs, although Osaka City University, Tokyo Metropolitan University, and Tsukuba University (starting in 1975) also offered doctoral work. Kyoto University became a major force in plant ecology through the doctorate after a garden for botanical ecology was established in 1964 and the university's premier chair in botany was renamed plant physiology and ecology at the same time. Hokkaido University's research unit in environmental science was founded in 1977, with doctoral work that included ecological administration. As of that year the postwar Education Ministry had supported new positions in botanical ecology at fifteen universities.[97] A sign of the importance now attached to ecological perspectives was the ministry's project on historical and cultural relations between people and nature in Japan from 2006 to 2011. Conducted by the National Research Institute on Humanity and Nature (Sōgō Chikyū Kankyōgaku Kenkyūjo), the ensuing 6-volume report edited by Yumoto Takakazu of Kyoto University drew on research by a team of more than 100 historians, archaeologists, and biologists and focused on the wise use of nature by people.[98] Although the doctrine of wise or sustainable use of natural resources was widely seen by conservationists as contrary to the outright protection of nature,[99] it was unlikely that a government-commissioned report would radically oppose established policies of national economic development; even to advocate sustainable use was a bold step in early twenty-first century Japan. The Yumoto report, and Japan's continuing priority on economic growth, indicates the limited impact of academic ecological perspectives on tangible public policy today. Nonetheless, despite the deficiencies in scientific training noted by Azuma Yoshirō and Iwatsuki Kunio at the beginning of this chapter, and despite the recent indifference of government agencies and private donors toward funding research in the life sciences, Japanese scholarship in plant biology operates at a very high level of achievement, building on a foundation of excellence by specialists in the Edo period and the pioneers of modern science in Japan nearly 150 years ago.

Afterword
Japanese Plant Biology and the World

"The rate of species destruction today is at about the level of 65 million years ago, when a major catastrophe, probably a huge asteroid, ended the age of the dinosaurs, opening the way for mammals to proliferate. The difference is that today we are the asteroid."[1] In these dramatic words, the linguist and environmental activist Noam Chomsky addressed the 2014 P.E.N. Club World Voices Festival in New York about the dangers human populations have posed to their natural surroundings since the invention of agriculture 12,000 years ago and especially since the industrial revolution beginning in the late eighteenth century. The epoch from the agricultural revolution to the present is known as the Holocene, but starting in 2000 some scholars led by the Nobel chemist Paul J. Crutzen have called the years since the industrial revolution the Anthropocene to reflect the heightened human impact on the planet.[2] While generally agreeing with Chomsky's outlook, neither the International Commission on Stratigraphy nor other groups of geologists have yet agreed on when, or whether, the Anthropocene epoch began.[3] Some concur with the Japanese plant biologist Hotta Mitsuru, writing in 1974, that environmental degradation started with early agricultural peoples who cleared forests for paddies, dryfields, fuel, fertilizer, and building materials[4]—prima facie evidence that the Anthropocene is coterminous with the Holocene epoch.

Such debates about periodization aside, Japanese plant biologists almost unanimously agree with colleagues in other countries about threats to flora from both human and nonhuman causes. Based on research done between 1964 and 1976, botanist Suzuki Sadao concluded that the number of species in the *Sasa* (bamboo grass) genus in Japan was fewer than previously thought, probably because of "ecological factors" such as climate change that caused certain species to be extirpated.[5] Biologist Yahara Tetsukazu in 1989 identified 35 native plant species as extinct, 147 as endangered, and 677 as vulnerable.[6] Using slightly different criteria, Iwatsuki Kunio, Yamazaki

Takashi, David E. Boufford, and Ōba Hideaki reported in 1995 that 300 plant species in Japan were considered threatened.[7] Five years later the Environment Agency (Kankyōchō) estimated that 24 percent of Japan's wild vascular plant species were in danger of extinction (*zetsumetsu no osore no aru shu*) and confirmed that 20 species were considered extinct.[8] The agency's study concluded that the plight of Japanese plant life "was graver than had been feared."[9] By 2006 Iwatsuki and colleagues had raised their estimate of threatened species to more than 1,000,[10] and in a 2015 report Yahara Tetsuzaku, Fujii Shinji, Itō Momomi, and Nagata Yoshio raised the number to 1,779.[11] Habitat loss and environmental changes menace plants around the globe; reclaiming wetlands and developing dryfields for housing, industry, and commerce are particularly severe factors in Japan, not to mention the enthusiasm of amateur horticulturalists for collecting plants from the wild.[12] At the same time, motivated in part to perpetuate plants under environmental threat, Japanese agencies and nurseries continue to exchange plants and seeds with counterparts in other countries.

Ecologist Nakamura Futoshi pointed out in 2015 that Japan's high diversity of species is partly attributable to the islands' "complex river morphology created by active volcanic and tectonic activities."[13] But biodiversity has been compromised inter alia by the country's 2,830 dams and the concretization of more than 95 percent of its river channels, both of which inhibit the growth of submerged vegetation and overhanging plants along riverbanks that shelter young fish from predators. According to one estimate 43 percent of aquatic plants needing at least temporary access to freshwater are endangered by agricultural pollution in rivers and floodplains.[14] By way of countermeasures, recent efforts to restore wetlands and river courses in Hokkaido's Kushiro and Shibetsu rivers have promoted a greater variety of flora and fauna; cutting back a check dam on the Iwaobetsu creek in Shiretoko national park by one meter has enhanced spawning habitats for salmon.[15] Alien plants that thrive along riverbanks, such as lance-leaved coreopsis (*Coreopsis lanceolata,* a tall wildflower in the sunflower family), originally introduced as an urban ornamental, crowd out existing plants and are considered invasive.[16]

Japan's public authorities continue to identify more and more such threats to native and naturalized species. The Invasive Alien Species Act of 2005 (Tokutei Gairai Seibutsu ni yoru Seitai Keitō ni kakaru Higai no Bōshi ni kansuru Hō) enumerated 112 plant and animal species as "specified foreign species," a list expanded and redefined in 2014 to include 424 "dangerous invasive species" for elimination.[17] One tenacious target was smooth cordgrass (*Spartina alterniflora,* smooth cordgrass or Atlantic cordgrass), a native of the Atlantic coast of the Americas that in Aichi and Kumamoto prefectures transforms tidal wetlands into grassy meadows, killing local grasses.[18] Underground root adaptations, rather than above-ground leaf traits and photosynthesis, may allow smooth cordgrass to grow efficiently and make it difficult to eradicate.[19] In short, Japan is a world leader in identifying plant species menaced by human-induced environmental factors but faces considerable fiscal hurdles and resistance from local communities in implementing effective protections.

Trees of all sorts may be the most evident targets of environmental threats to Japanese plant life. Globally about 10 percent of the 80,000 known tree species are threatened with extinction; of these, 18 species exist only in cultivation.[20] Forest trees in Japan number about 600 species and are endangered by environmental factors such as pathogens, fragmentation, herds of deer, and global warming, as well as locally excessive logging, overhunting, use of pesticides, air pollution, and acid rain.[21] Sulfurous gases from refining ores and poisonous sludges produced by paper mills degrade the environment and are particular hazards to tree populations. One example of excessive logging is beech trees. Although less commercially valuable than oak or maple, some beech species are widely used in Japan for plywood and flooring after being clear-cut by bulldozers and pulled from steep forest slopes by wire cables.[22] Although Japan is often called a country of "tree culture" (*ki no bunka*), today fewer houses are built of wood for reasons of economy, concern about weathering, and consumer preference for earthquake-proof concrete and steel. Japan supplies about 28 percent of its own timber needs and imports the rest, chiefly from Australia, Canada, and the United States; Malaysia and Indonesia together provide 9.2 percent.[23] One result of such extensive importation is that ordinary Japanese have "a remarkably low appreciation of [Japanese] forest areas,"[24] prompting a government-industry campaign since 2004 to promote "forest learning" (*mokuiku*) to "nurture feelings for trees and forests" and to "link people to forests." In recent years Japanese media, concerned about global warming and carbon emissions, have touted the slogan "protect the forests" (*shinrin o mamore*)[25] for reasons of conservation rather than commerce.

The effects of warmer temperatures are particularly evident in early flowering cherries. A study at Tama Forest Science Garden found that trees of roughly the same age from 17 cherry species and hybrids flowered an average of 5.5 days earlier in 2005 than in 1981, during which average ambient temperatures at the garden increased a fraction over 1 degree centigrade. An interesting finding is that earlier flowering seems to lead to longer flowering.[26] To aid certain tree species under threat, scientists have debated for more than two decades whether to move them to new habitats, a proposal called "assisted migration." Skeptics find it problematic to interfere with how wild species evolve and fear that the assisted species could become pests to local vegetation in the new locale. Proponents believe that moving trees north could provide northern trees the genetic resources to survive.[27] A Japanese study in 2010 projected that climate change could raise lower-altitude vegetation ranges upward by 293 meters by 2100. Alpine, subalpine, and sub-boreal species could be moved by assisted migration to Hokkaido, where they currently do not exist, but such a strategy "may entail the risk of introducing biotic competition or interbreeding."[28]

Nikkō is a major tourist venue centering on the Tōshōgū shrine that memorializes the first Tokugawa shogun, Ieyasu (1543–1616, r. 1603–1605). A number of the town's buildings are designated World Heritage sites, 9 are considered national treasures, and 94 are listed as important cultural properties. A heavy winter storm felled more than 100 trees in the area in February 2014, prompting a census that identified 80 cryptomerias and cypresses as being at risk of falling on temples and shrines because of rotten trunks. Forty of the 80 were deemed so rotten that they might collapse imminently, so officials began taking them down.[29] That same year Boy Scouts surveyed the Miho no Matsubara pine forest in Shimizu, Shizuoka city, and calculated that the number of trees had declined by more than one third from the 1989 estimate of 54,000, mainly, scholars think, because of weevil infestations and infectious diseases.[30] An even more drastic decline afflicted the Chichibu birch (*Betula chichibuensis*), which grows wild only in the Chichibu mountains northwest of Tokyo and is considered critically endangered, even though today 24 individuals of this tree have been counted, up from 21 in the mid-1990s. Botanists in Great Britain have sprouted about 100 Chichibu birches from seeds collected in the mountains and planted in compost, a step toward conservation of the species.[31] Oak beetles, present in Japan since the Edo period, have spread to every region of Honshu since 1990, devouring young trees thinner than ten centimeters in diameter. Dying oaks and red pines receive little attention from government foresters, allowing diseases to spread to healthy trees. Yet local residents often oppose using chemical agents to forestall plant diseases, even when trees begin to die.[32]

Of particular concern to biologists are the effects of the March 2011 meltdown at Fukushima Daiichi nuclear power station on the region's vegetation. A survey in 2015 found that stunting, bifurcation, and other morphological changes affected 90 percent of fir trees within 3.5 kilometers of the meltdown, 40 percent of those in a sample 8.5 kilometers distant, and 30 percent at a site 15 kilometers away.[33] Yet specific genetic mutations in the trees cannot yet, and perhaps ever, be established. Chomsky's warning about environmental damage arose in part when this power station built to serve human needs during Japan's age of high industrialization was swamped by a natural event originating 70 kilometers offshore, evidence to some that humanity had long since entered the Anthropocene epoch.

These and many other incidents of damage to or attenuation of tree species helped prompt the government's Forestry Agency (Rin'yachō) in 2014 to start cutting down more than 5 million hectares of mature plantation forests, planting seedlings as replacements. One obstacle was identifying the unknown owners of 1 million hectares of forests in private hands. Heirs failed to change ownership registrations, owners abandoned their properties and moved to cities, and still others were put off by the ¥500,000 fees to change ownership records for lands often worth less than the cost of recording the change.[34] In May 2018 the government enacted a Forestry Management Control Law that made it easier to sell state-owned forests to private investors, as also happened when the Main Crop Seeds Law of 1952 was abolished in April 2018, allowing private companies to enter the seed business. Critics attacked both actions for enabling the sale of public assets to the private sector,[35] reminiscent of the Meiji government's selloff of state-owned pilot industries to financial trusts (*zaibatsu*) during the Matsukata deflation of 1881–1886. In short, privatization may be the future of both national environmental policy and botanical research, forcing activists and plant biologists to seek nonstate sources of support in order to protect Japan's ecosystems and continue scholarly investigations of the country's rich endowment of flora.

The main findings and interpretive patterns of *Land of Plants in Motion: Japanese Botany and the World* may be summarized as follows:

1. The ancestors of most plants now alive in Japan arrived there from China and the eastern Himalaya during Tertiary times (66–2.6 million years ago). The forebears of many Japanese plants also spread from the East Asian floristic zone to eastern North America by 5 million years ago, by way of plant interchanges among East Asia, Europe, and North America via land bridges across the North Atlantic and the Bering sea. The limited damage caused in Japan by global glacial cycles 110,000 to 11,700 years ago helps account for the richness of Japanese flora since. Both paleoclimatological processes and modern commercial motives explain the movement of plants across time and around the world. About 1,860 current terrestrial species are considered native to Japan, including those such as bamboo that have been in Japan so long that they are regarded as indigenous. The majority of plant species seen as "Japanese" elsewhere in the world were introduced from Japan during the past two centuries, mainly for ornamental purposes but also for scientific study.
2. Early modern (1600–1868) scientists in Japan drew on Chinese knowledge of herbal medicine but followed paths specific to Japan's natural

setting, achieving sophisticated insights into plant life through empirical scientific understandings commensurate with but distinct from their European counterparts. Employees of the Dutch East India Company at Nagasaki helped spread knowledge of Japanese plant life to Europe through their writings and botanical collections. By the early nineteenth century Japanese specialists gathered specimens, conducted surveys, drew images, and described and classified plants in detail—tantamount to botany in method if not yet analytical technique. A handful of scholars set aside Chinese herbalism and conventional natural history to introduce Western studies of plant morphology, physiology, and chemistry under the new rubric of *shokugaku* (botany, later *shokubutsugaku*).

3. Missionary reports, accounts by distant travelers, and a fascination with natural history helped fuel an appetite for plants considered exotic by early modern Europeans and nineteenth-century North Americans. Philipp Franz von Siebold returned from Japan to the Netherlands in 1830 to become Europe's foremost authority on Japanese flora. Plant collectors from botanical gardens and commercial nurseries in Western Europe and the United States streamed to East Asia in the later nineteenth and early twentieth centuries, sending home thousands of Japanese herbarium specimens for scientific research and both seeds and live plants from Japan for horticulturalists, who by now have indigenized their descendants to the point that they are rarely considered foreign imports today. By the early twentieth century nursery businesses in Yokohama became the chief providers of Japanese plants to the rest of the world.
4. Japan's extensive encounter with Western culture after 1868 soon led its plant specialists to master European principles of taxonomy and morphology, so that botany became a full-fledged discipline starting in the late 1870s. The University of Tokyo, its Koishikawa Botanical Gardens, and the Botanical Society of Tokyo became the pivot of plant biology research in Japan for the next generation or more. The pioneering botanist Yatabe Ryōkichi established difference, if not quite independence, from Western botanists when he declared in 1890 that "Japanese plants should be given scientific names by Japanese."[36] The self-educated scholar Makino Tomitarō never studied overseas, yet he became Japan's best-known plant biologist abroad. In 1896 researchers at Tokyo Imperial University made significant discoveries of swimming sperm cells in two seed-bearing trees, without scholarly assistance from Europe or America. But Japanese specialists also sought validation from the world's scholarly community through publishing their research in English-language journals of high quality, and by the early twentieth century

Tokyo became a magnet for botanical researchers from other countries. At that point Japanese plant biologists had developed a wide range of research specialties well beyond the morphology and taxonomy introduced forty years earlier.

5. Makino Tomitarō and Minakata Kumagusu, two plant biologists from unconventional backgrounds, were exceptionally productive scholars with wide impact on fellow scientists and amateur botanists in the first half of the twentieth century. Makino pioneered the field of public (amateur) botany, drawing on popular fascination with plant life in Japan since the Edo period. Both Makino and Minakata identified rare species and opened up new fields of research in plant biology. Their work paralleled genera- or species-specific research by other scholars as appointments in plant biology gradually were made at a number of universities and research institutes beyond Tokyo and Sapporo. Botanical studies in Japan between the world wars reached high levels of empirical detail and methodological finesse comparable with scholarship in Europe and North America.
6. Japan's age of empire, 1895–1945, was one of plant collection rather than exploration, intended for both scientific and economic benefits. The effect was to make Japan's empire of colonization and military conquest visible at home through direct engagement with vegetation from lands under Japanese domination, both in research laboratories and at botanical gardens open to the public, especially those displaying exotic tropical plants. In this way a visual and material culture of imperial hegemony emerged in Japan during the 1930s and early 1940s. Japanese plant researchers scoured territories from Karafuto (Sakhalin) in the north through Korea, northeast China, and Taiwan to the tropics of Southeast Asia and Pacific islands for species unfamiliar at home. Censuses of forest products, crops, and other economically important plants helped leaders at home know their empire by making it legible. After World War Two many of Japan's overseas specialists brought the botany of empire home by setting up laboratories at both established and newly founded universities and research institutes, cementing Japan's prominence in studying the botany of Asia that continues today.
7. Graduate training for botanical research after 1945 continued the prewar pattern of apprenticeships to senior professors, despite repeated criticisms of the common system of choosing successors to departing university professors from among their own students, with the result that preparation for research was different from training in most other countries, but not necessarily better or worse. The infrastructure for research expanded with the development of new graduate programs, publication of new

scholarly journals in specialized fields of plant biology, and growth of strong scientific networks among specialists throughout the country, centering on University of Tokyo. New scholarly societies and revivified older ones brought together researchers in various biological subfields and encouraged outreach to counterparts abroad, setting both Japanese plants and studies of them in motion around the globe. New botanical gardens, many with greenhouses, aided research as well as horticultural appreciation in localities distant from major scholarly centers in Sapporo, Sendai, Tsukuba, Tokyo, Kyoto, and Osaka. Plant physiology thrived in Japan during the late twentieth century, while microbiological research focused increasingly on biochemical genetics. Molecular biology became a mainstream field of Japanese plant biology in the 1970s, as did recombinant DNA research starting in the 1980s. Regional studies of plant life in Japan have proliferated, aided by amateur researchers familiar with local vegetation. Plant ecology became a widely recognized field in the later twentieth century and remains a leading area of study within environmental science today. Basic research in all scientific fields was hard hit by a 5 percent reduction in government funding between 2012 and 2017, ameliorated only slightly since and hardly offset by significant private support. Despite any narrowness in graduate training and recent setbacks in funding, Japanese plant biologists enjoy international recognition in many subfields of research, continuing a pattern of excellence set by Edo-period specialists and pioneers of modern science in Japan since the early Meiji era.

Viewed through both the hyperopic prism of geological time and the myopic lenses of human history, the experiences of flora in Japan—like those elsewhere on the planet—suggest that plants will continue in motion far into the future. As in past epochs, warming climates can be expected to induce many Japanese species to move laterally inland, vertically upslope, and horizontally northward in search of terrain compatible with their ecological needs. Crops and other vegetation that cannot easily reposition themselves may face tough going as ambient temperatures rise. The visual landscape of the Japanese islands, now abundantly green despite a substantial built environment, seems likely to be considerably altered across distant or not so distant time, possibly beyond recognition to persons now alive. This sobering thought should heighten humanity's respect for the tenacity and hardiness of plant life in Japan and elsewhere on the globe. Plants have existed on Earth for hundreds of millions of years, survived several major extinction events, and are likely to overcome all but the most cataclysmic environmental perils posed by the future.

Notes

Preface

1. Crown Prince Naruhito, speech, Tokyo, November 18, 2018, quoted in *Mainichi shinbun* online, November 19, 2018.

Introduction

1. John L. Creech to Tsukamoto Yōtarō, March 9, 1994, in Tsukamoto and Creech, 179.
2. Ibid., 168.
3. Hodgson, 319.
4. Hotta Mitsuru, 128, 138–139, 146; Suka, Okamoto, and Ushimaru, 7; Friis, Crane, and Pedersen, 95.
5. Merrill, 37. See also Ōi, 1.
6. Nihon Shokubutsu Gakkai, *Hyakka,* 9; Hara, *Catalogue,* 10, 13. Hara's surveys included plants in Nepal, Sikkim, and Bhutan.
7. Boufford and Ohba, v.
8. Levy-Yamamori and Taaffe, 19.
9. Spongberg, 88; *Japan Today* online, March 31, 2015.
10. Bartlett and Shohara, 73; Levy-Yamamori and Taaffe, 19, 37; Merrill, 37; Tsukamoto and Creech, 86–90, 100–101, 127; Nihon Shokubutsu Gakkai, *Hyakunen,* 85; Tamura, 27–29.
11. Iwashina and Ebihara, iii. For other estimates, see Levy-Yamamori and Taaffe, 19; Tsukamoto and Creech, 167.
12. Tsukamoto and Creech, 167.
13. See Miyawaki, Iwatsuki, and Grandtner, 4, 61; Miyawaki.
14. Yokohama Ueki Kabushiki Kaisha 1901, 78.
15. Elliott, 21–22; Spongberg, 227.
16. Schiebinger and Swan, 2.
17. Matsui and Iwasaki, 10–11.
18. Shibata.
19. Nihon Shokubutsu Gakkai, *Hyakunen,* 117, 120, 154–155.
20. Ibid., 197–198. For environmental science terms, see Parsons, "Wildness," 659.
21. Kokuritsu Kagaku Hakubutsukan Tsukuba, 30.
22. Iwatsuki Kunio, 87.
23. Ōba, *Nihon no zetsumetsu,* xvi.
24. Woese and Fox, 5088–5090.

25. Quammen.

26. International Botanical Congress (2012), ix.

27. International Botanical Congress (2006), 1–2.

28. Jones, ix–x. Jones notes that much archaeobotany before the 1990s was based on German research.

29. Madella, Lancelotti, and Savard, 3.

30. Kira, 230.

31. Ibid., 232.

32. Smith, 84, 89; Novacek, SR7. Insects account for the majority of living species.

33. "Tsukuba Botanical Garden," National Museum of Nature and Science, accessed August 28, 2019, www.tbg.kahaku.go.jp. See Goldenberg; Horikawa, *Atlas,* 1:9; Fleming. Horikawa enumerates 600 fern, 1,500 moss and other nonvascular, and more than 1,000 lichen species in Japan. Fleming notes that the biomass of land plants is a thousand times that of animals.

34. Creech, *Ornamental,* 2.

35. Iwatsuki Kunio, 148, 152.

36. Takeda, *Systematic,* 3. As early as 1891 the taxonomist Tanaka Yoshio (1838–1916) distinguished plants that were useful (largely foods) from those characterized as economic (mainly supplies and materials). Tanaka Yoshio, 1:1–35, 1:79–88.

37. Kanai, 1; Takeda, *Alpine,* 1–13.

38. Takeda, *Alpine,* 10.

39. Hotta Mitsuru, 8–13. See Miyoshi and Makino, *Pocket-Atlas,* ii.

40. Takeda, *Alpine,* 21–24.

41. Ibid., 14–29.

42. Levy-Yamamori and Taaffe, 28. The authors point out that both the Shirakami beech (*Fagus crenata*) forest and the *Cryptomeria japonica* forest on Yakushima island were named Unesco World Heritage sites in 1993.

43. Hotta Mitsuru, 41.

44. Ibid., 42–44; see also 30–36, 132–135.

45. Ibid., 71. Hotta's frank assessment of forest vegetation contradicts advocates of "forest culture" such as Umehara Takeshi, Yasuda Yoshinori, and other "reactionary ecologists" who believe forests exemplify Japan's aesthetic ties to nature. See Reitan.

46. Levy-Yamamori and Taaffe, 25.

47. *Asahi* online, March 13, 2015.

48. Crane, 90–91, 239.

49. Hotta Mitsuru, 54–61, 64, 86–88.

50. Levy-Yamamori and Taaffe, 17–25.

51. Hotta Mitsuru, 77–84, 136–137.

52. Pollan, 94–95. On page 92 Pollan summarizes the controversial field of plant neurobiology, which posits that plants have intelligence and may be able to feel emotions. See Chung and Williams, 26–27, on communication among trees.

53. Seki et al., 332.

54. Darwin.

55. Parmesan and Yohe, 37–42.

56. Neilson et al., 750.

57. Iwashina and Ebihara, iii. In 1965 Suzuki Shigetaka estimated that 2,800 plant species "belong only to Japan," including a number of naturalized taxa. Suzuki Shigetaka, 2. Levy-Yamamori and Taaffe count about 1,867 endemic species of flowering plants plus 2,500

to 3,000 native moss species. Levy-Yamamori and Taaffe, 7, 403. Of Hawai'i's 1,200 native plant species, 90 percent are thought to exist nowhere else. *New York Times,* May 25, 2018, A3.

58. Hotta Mitsuru, 4, 42–44: Ebihara, "Nihon," 140. See Levy-Yamamori and Taaffe, 7.

59. Ebihara, "Seibutsu," 137–139.

60. "Tsukuba Botanical Garden," National Museum of Nature and Science, accessed August 28, 2019, www.tbg.kahaku.go.jp.

61. Fairchild; Tsukamoto and Creech, 152.

62. Brown, 8. See Creech, "Foreword," 10; Creech, *Ornamental,* 1.

63. Jonas, "Japanese Inspiration," 4.

64. Anthony Ricciardi, quoted in Kolbert, 198.

65. Forseth and Innis, 401.

66. Li et al., 381–382; Leicht-Young, 7–8.

67. Li et al., 389. *Harper's Magazine,* October 2014, 11, placed the annual cost of invasive animal, plant, and microbe species to the U.S. economy at $120 billion.

68. Lexington.

69. Cockburn, 58. See Klinkenborg, SR10.

70. Mark Davis, quoted in Cockburn, 59.

71. Stephen Jay Gould, quoted in Cockburn, 58.

72. Peter Del Tredici, quoted in Cockburn, 59.

73. David Boufford, personal communication, June 16, 2015.

74. Manke.

75. *Mainichi* online, November 11, 2017.

76. Lexington; Li et al., 382; Jonas, "Plants," 68; Schierenbeck, 391. Japanese honeysuckle was collected by Carl Peter Thunberg at Nagasaki as early as 1784.

77. Levy-Yamamori and Taaffe, 22–23; Forseth and Innis, 402–413; Leicht-Young, 8–10. Kudzu (Jp. *kuzu*) is a homophone of *kuzu* (rubbish) but written with a different character.

78. *Mainichi* online, November 11, 2017; Watanabe, 9, 23, 37, 65, 100; Hollingworth; Knight; Nelson, 65–66. Similar to knotweed is *Fallopia sachalinensis* (giant knotweed, previously *Polygonum sachalinensis*), which grows four meters tall. Other parasitic insects from Japan include emerald ash borers (*Agrilus planipennis*), accidentally introduced to North America in the 1990s, and Japanese beetles (*Popillia japonica*), not considered a pest of much economic significance in Japan and first found in the United States in 1916. *Boston Globe* online, August 1, 2014; Clausen, 1–3, 44.

79. Ōba, *Nihon no kika,* xix. Phytogeographer Tatewaki Misao (1899–1976) conducted extensive studies of Hokkaido's vegetation, noting that 10 percent of its 1,800 species of higher plants were considered endemic and a small number primeval. Tatewaki, 1.

80. Morita, ii, 3–7. Hotta Mitsuru, 112–117, provides extensive lists of native plants in Japan. Big real estate companies and the Environment Ministry are jointly promoting gardens featuring native plants and animals at new office building sites. *Mainichi* online, July 27, 2014.

81. Nihon Shokubutsu Gakkai, *Hyakunen,* 708; Morita, 4. For earlier estimates, see Ōba, *Nihon no kika,* vi, xix; Shimizu, 14.

82. Shimizu, 19–20; Shizen, 5.

83. Nihon Shokubutsu Gakkai, *Hyakka,* 709; Morita, 9.

84. Shimizu, 26.

85. Shizen, 21–28; Shimizu, 26; Nihon Shokubutsu Gakkai, *Hyakka,* 708–709. The full name of the Invasive Alien Species Law is Tokutei Gairai Seibutsu ni yoru Seitai Keitō ni Kakaru Higai no Bōshi ni Kansuru Hōritsu.

86. Nihon Shokubutsu Gakkai, *Hyakka*, 742. By contrast the "beech-withering disease" is a domestic Japanese fungus caused by oak wilt (*Raffaela quercivora*). Despite its common name in Japanese, this fungus attacks oaks and chinquapins but not beeches (ibid.).

87. Brombert, 1.

Chapter 1: East Asia's Plants in Geological Time

1. *New York Times*, January 22, 2019, D3.

2. Komiya Tsuyoshi, quoted in *Asahi* online, September 28, 2017. Komiya's findings were published in *Nature* online, September 27, 2017. In June 2018 the American Space Agency NASA announced that the Mars rover *Curiosity*, drilling into rocks 3.5 billion years old, had detected carbon molecules that could be building blocks of living organisms on that planet. See Voosen; *New York Times*, June 8, 2018, A12.

3. Kokuritsu Kagaku Hakubutsukan Tsukuba, 31–32, 44. Mosses survived because they developed a cuticula layer of skin to prevent drying and flavonoids to protect against ultraviolet light. Ibid., 32.

4. Tamura, 1–4.

5. Ibid., 102, 112–115; Crane, 87. Ginkgos are considered gymnosperms but have swimming sperm cells like early ferns and cycads. Spongberg, 86.

6. Tamura, 5. For an earlier study of Japanese vegetation in geological time, see Numata.

7. Tamura, 115; Levy-Yamamori and Taaffe, 28; Crane, 61; Spongberg, 86–88.

8. Manchester, 472–475; Crane, 5, 152–157.

9. Manchester, 475–476.

10. Crane, 6. Crane cautions that the nominally wild stands of ginkgos found in China today may have been planted by people several centuries ago (ibid., 162).

11. "A-bombed Trees in Hiroshima, Japan," Kwanten, accessed August 28, 2019, https://kwanten.home.

12. "Ginkgo 'Kitakanegasawa no Ichou' Near the Border of the Kita Kanegasawa in Nishitsugaru-gun, Fukaura-machi, Tohoku, Japan," Monumental Trees, accessed August 28, 2019, https://www.monumentaltrees.com.

13. Novacek, SR7.

14. *Asahi* online, October 18, 2014; Kolbert, 96–97; Crane, 254–255; Brannen, SR2.

15. Brannen, SR2; Crane, 255.

16. Kokuritsu Kagaku Hakubutsukan Tsukuba, 33.

17. Tamura, 32.

18. Ibid., 28, 117; Hotta Mitsuru, 103.

19. Novacek, SR7.

20. Askins, 8–11.

21. Tamura, 48, 53, 77, 80.

22. Crane, 123. Crane adds, "It is less clear that major groups of plants were lost at a global scale."

23. Askins, 18.

24. Ibid., 12–19, 31; Tamura, 89–90.

25. Friis, Crane, and Pedersen, 1.

26. Hotta Mitsuru, 99.

27. Friis, Crane, and Pedersen, ix, 95, 475, 498; Tamura, 122. Friis, Crane, and Pedersen, 2–3, provide two possible cladograms to clarify angiosperm evolution, based on advances in phylogenetic systematics since 1980.

28. Friis, Crane, and Pedersen, 3–6, 275; Tamura, 84–88.
29. Tamura, 34–37, 56.
30. Ibid., 45, 72.
31. Hotta Mitsuru, 104–105.
32. Askins, 11.
33. Chaney, 3. See Tiffney, 86–87; Tanai, 235.
34. Askins, 20; Hotta Mitsuru, 122–123.
35. Hara, *Distribution,* 2:89.
36. Tamura, 69, 73–76, 81.
37. Manchester et al., 1, 31–33; Tiffney, 83. A plant group that has not dispersed widely is known as an example of neoendemism.
38. Tiffney, 83–84.
39. Askins, 21.
40. Hotta Mitsuru, 107, 128–131.
41. *Asahi* online, July 28, 2015.
42. Hotta Mitsuru, 127–128, 146.
43. Ibid., 139.
44. Crane, 157; Manchester et al., 508–509. Manchester et al., 477–482, lists numerous plants found in North America in the Tertiary period that now exist only in East Asia.
45. Askins, 22. European forests in the late Tertiary period were less diverse, dominated by conifers.
46. Hara, *Distribution,* 2:93. See Hara, "Patterns," 55, 59.
47. Hotta Mitsuru, 107, 140. In 2017 Japanese researchers sought to rename the Middle Pleistocene epoch (770,000–126,000 years ago) as Chibanian, based on evidence of geomagnetic reversal found in a stratum along the Yōrō river in Chiba. Their application to the International Union of Geological Sciences was later thrown into doubt by charges of fabricated data but finally approved on January 17, 2020. *Asahi shinbun* online, April 15, 2016; *Mainichi shinbun* online, May 6, 2017, May 12, 2018, January 17, 2020.
48. Suka, Okamoto, and Ushimaru, 52–54. Askins, 23, enumerates 18 to 20 ice cycles throughout the Pleistocene epoch.
49. Horikawa, *Atlas,* 1:8; Nitzelius, 3.
50. Cf. Abe Kōbō, *Daiyon kanpyōki* (Tokyo: Kōdansha, 1958–1959), translated as *Inter Ice Age Four* by E. Dale Saunders (New York: Knopf, 1970).
51. Yano, 208; Hotta Mitsuru, 107, 142–144; Suka, Okamoto, and Ushimaru, 53.
52. Miki, "Plant Fossils," 337–338; Hara, *Distribution,* 2:92.
53. Hara, *Distribution,* 2:92; Tamura, 93.
54. Spongberg, 217; Yano, 210.
55. Manchester et al., 1.
56. Crawford, 15.
57. Suka, Okamoto, and Ushimaru, 15; Nihon Shokubutsu Gakkai, *Hyakka,* 749. Today *satoyama* means restoring rural lands on forest margins. See Havens, *Parkscapes,* 167–168. Jōmon *satoyama* were natural environments, whereas today's are entirely constructed by humans.
58. Yano, 214.
59. Yano, 210–214; Suka, Okamoto, and Ushimaru, 12–19.
60. Hotta Mitsuru, 146–147.
61. Suka, Okamoto, and Ushimaru, 36, 43–44.
62. Miyashita-Theado, 13–17. For literary criticism of "Musashino," see Thornber, *Ecoambiguity,* 481n197.

63. Suka, Okamoto, and Ushimaru, 30; see 4–7, 45–46.

64. On reforestation, see Totman, *Green,* and Totman, *Japan's Imperial Forest.*

65. Suka, Okamoto, and Ushimaru, 33; see 14, 28–29; Nihon Shokubutsu Gakkai, *Hyakka,* 749.

66. Hara, *Distribution,* 2:91–92.

67. For the Ōkunitama shrine, see Hardacre, 412–413, 485–488, 493, 500.

68. Lewis and Wigen.

69. Angier, D1, D3.

70. "Dance of the Giant Continents," Burke Museum, University of Washington, Washington State Museum of History and Culture, accessed August 28, 2019, http://www.burkemuseum.org.

71. Nance and Murphy.

72. "What Is Gondwana?" Live Science.com, part of Future U.S., accessed August 28, 2019, https://www.livescience.com.

73. Chwastyk, 30–31.

74. Milne, 466; Manchester et al., 33; Askins, 5; Gheerbrant and Rage, 224–246. Manchester, writing in 1999, dates the end of the North Atlantic land bridges to between 38 and 34 million years ago. Manchester, 513.

75. Scharff, 1–28.

76. Milne, 465; Tiffney, 88; Ager, 19–32; Hsu, 490.

77. Parsons, "Wildness," 654.

78. Wen, 421; Parsons, "Natural," 37, 68. Parsons reports on page 68 that in 1841 the eastern United States produced 64 million ginseng roots. See Kuriyama, 61–72.

79. Boufford and Spongberg, 421; Graham, "Outline," 1; Boufford and Spongberg, 424.

80. Thunberg, *Flora.*

81. Boufford and Spongberg, 424–425. Sam Bass Warner, Jr., points out that botany in the United States focused on agriculture before the nineteenth century. The Massachusetts Horticultural Society was founded in 1829, the first of its kind in the United States. Warner, ix.

82. Boufford and Spongberg, 425; Nuttall.

83. Siebold, *Flora.*

84. Gray, "Account," 2:305–332. Thirteen new species were *Carices* (a genus of grassy plants); also included in the collection were thirty algae species. See Graham, "Outline," 17; Tsukamoto and Creech, 168–169. Williams was a childhood friend of Gray.

85. Gray, "Account," 2:305. For plant collecting by the Perry expedition, see Koyama, *Plant Collecting;* for drawings of the species brought back by Perry's expedition and by the North Pacific expedition of 1855–1856, see United States Naval Expedition.

86. Wright.

87. Del Tredici, "Introduction."

88. Gray, "Diagnostic," 437, 445–447; Gray, "Extract," 187; Hotta Mitsuru, 119. See Barnes, 118. Spongberg, 147–148, notes that historian and horticulturalist Francis Parkman (1823–1893) grew Kousa dogwoods (*Cornus kousa*) and American flowering dogwoods (*Cornus florida L.*) at his Brookline, Massachusetts, estate, yielding proof of two closely related species that were dispersed by monkeys in Japan and birds in North America.

89. Gray, "Diagnostic"; Charles Sprague Sargent (1888), quoted in Sutton, 198.

90. Boufford and Spongberg, 428, 431.

91. Sargent, 8. Sutton, 206, notes that Sargent's *Forest Flora* was not systematic but instead mostly personal observations of Japan.

92. Ibid., 431.

93. Utech, 189.

94. Tiffney, 73–74; Spongberg, 143–145. Hotta Mitsuru, 119, largely supports Gray's theory of Arctic ice advancing southward.

95. Donoghue and Smith, 1633.

96. Milne, 466.

97. Boufford, "Eastern," 51.

98. Manchester, 501. Hsu, 490, likewise found that eastern North American plants similar to those in China migrated from East Asia via the Atlantic as recently as 40 million years ago.

99. Manchester, 512.

100. Sharp, 62.

101. Wen, 439–445.

102. Tiffney, 89.

103. See ibid., 77–79, 85–86.

104. Ibid., 85.

105. Manchester, 499, further classifies Tertiary plants into (1) species shared between North America and East Asia, and (2) those endemic to North America and not found elsewhere.

106. Boufford, "Eastern," 49–50.

107. Kudō, "Yezo," 291. Kudō also found that 45.2 percent of Hokkaido species were shared with Korea, 39.5 percent with China, and 27.9 percent with Europe. Some common species were likely the result of exports to Europe and North America in recent centuries.

108. Hara, *Observationes,* 20, 640.

109. Hara, *List.*

110. Iwatsuki Zennosuke, 245–248.

111. Hotta Mitsuru, 119.

112. Miyawaki, Iwatsuki, and Grandtner, 4.

113. Wu, 577.

114. Tsukamoto and Creech, 35. *Vicariads* are seemingly related pairs in separate environments that have grown morphologically and genetically distinct from a common ancestor. Dividing plant populations into separate subpopulations via continental fragmentation is more likely to account for most instances of vicariance than seed dispersal by migratory birds. See Milne, 469–471.

115. Wen, 422–423, 428–432.

116. Askins, ix, 3–5; Tamura, 24; H. Li.

117. Wen, 421, 427, 436–438. Wen, 423, notes that molecular, fossil, and breeding data suggest morphological stasis is common for some disjunct plants in the two regions.

118. Donoghue and Smith, 1633; Askins, 3; Wen, 434.

119. Wilson, *Conifers,* v–vii.

120. Hotta Mitsuru, 121. Graham, *Floristics,* 93, noted in 1972 that more and more similarities of vegetation in the western United States with that in Japan were being discovered.

121. Wen, 435, 445.

Chapter 2: Plants in Early Modern Japan

1. Shirai, *Shokubutsu;* Shirai, "Brief," 213. See Bartlett and Shohara, 17–19.

2. *Mainichi shinbun* online, April 10, 2014. In August 2018 the Japanese government announced plans to use big data (about one million recipients per medication) to study the

efficacy and side effects of Chinese herbal medicine, using a data server at Keiō University Hospital's Center for Kanpō Medicine in Tokyo. *Yomiuri shinbun* online, August 25, 2018.

3. The Medicinal Plant Garden at Kumamoto University contains more than one thousand species. "Medicinal Plant Garden," Kumamoto University, accessed August 28, 2019, http://www.pharm.kumamoto-u.ac.jp. See Crane, 242–244; Ichimura, I; Kimura Kōichi, I; on Ainu plants, see Batchelor and Miyabe, 200–213.

4. Goble, 47; Merrill and Walker, 551; Shirai, "Brief," 214; Itoigawa, 2–3; Nihon Shokubutsu Gakkai, *Hyakka,* 7.

5. Lock, 24, 50–51.

6. Ōnuki.

7. Seaton et al., 76.

8. Shirai, "Brief," 214; Takemi, 1–2; Walker, 74; Kihara, 158; Nihon Shokubutsu Gakkai, *Hyakka,* 7; Muntschick, 76.

9. Cook, 26, 32; Kimura Kōichi, ii.

10. Marcon, *Knowledge*, 7, 28–43, 50; Hanson, 77; Nappi, 4–7.

11. Nappi, 20; Marcon, *Knowledge*, 52, 55; Kihara, 160.

12. Cook, 339–341.

13. Nihon Shokubutsu Gakkai, *Hyakka*, 2–3.

14. Warner, vii–viii.

15. Muntschick, 71–72; Williams, 4.

16. Kimura and Kitamura, 16; Nelson, 52. For smuggling, see Thai.

17. Kimura and Kitamura, 16; Muntschick, 75; Tsukamoto and Creech, 132; "'Dwarf Trees' from George Meister's *Der Orientalish-Indianische Kunst und Lust-Gärtner,*" Magical Miniature Landscapes, accessed August 28, 2019, http://www.magiminiland.org. Meister's book, published at Dresden by J. Riedel in 1692, contained 338 pages and 16 leaves of plates. Numerous copies of the first edition can be found in the United States, Germany, and elsewhere; the book has been reprinted several times, most recently in 1973.

18. Kaempfer, *History*, 1:183.

19. Engelbert Kaempfer, quoted in Bowers, 46. See Stearn, "Kaempfer," 103–104, 108; Herries, 1; Crane, 204, 209.

20. Muntschick, 77; Spongberg, 84.

21. Hirohito, "Linné," 116; Muntschick, 79–81; Stearn, 111; Kowenhoven and Forrer, 76.

22. Muntschick, 85, 94; Levy-Yamamori and Taaffe, 32.

23. Engelbert Kaempfer, *History,* quoted in Kaempfer, *Kaempfer's Japan*, 69. See Kaempfer, *History*, 1:177–183.

24. Kaempfer, *History*, 3:215–248.

25. Rietbergen, 212.

26. Itoigawa, 1–2.

27. Davis, 106.

28. Ogilvie, 269–270.

29. Marcon, *Knowledge*, 86.

30. Nihon Shokubutsu Gakkai, *Hyakka*, 7; Marcon, *Knowledge*, 77–78, 91–94; Kihara, 161; Trambaiolo, 309; Itō Tokutarō, "History," 1; Haga, 245.

31. Fukui, 45–53.

32. Marcon, "Inventorying," 191–196; Marcon, *Knowledge*, 103–105, 109–110, 124.

33. Bartlett and Shohara, 49.

34. Itoigawa, 4–5.

35. Marcon, *Knowledge*, 114; Parsons, "Natural," 37, 68.

36. Kasaya, 173–176; Trambaiolo, 310.

37. Lock, 134–135; Kasaya, 173.

38. Goodman, 60. See Bartlett and Shohara, 27; Marcon, *Knowledge*, 114, 117; Nihon Shokubutsu Gakkai, *Hyakka*, 22; Nihon Shokubutsuen Kyōkai, 280.

39. Kawakami Sachio, 5–7; Havens, *Parkscapes*, 134–135; Iwatsuki Kunio, *Nihon*, 10, 102–103.

40. Scott.

41. Shirai, "Brief," 220; Marcon, "Inventorying," 190, 198.

42. Marcon, "Inventorying," 203; Sagers, 45–52.

43. Kasaya, 168, 181–183; Haga, 242.

44. Shirahata, 264, 269–272; Walravens, 7–10.

45. See Ueno Masuzō, *Jidai*, 173; Marcon, *Knowledge*, 5–6.

46. *Kodansha*, 7:334; Ueno Masuzō, *Jidai*, 151–154; *Kodansha*, 3:142.

47. Haga, 248; see also 247–249.

48. Bartlett and Shohara, 59. See Ōtsuka and Yakazu.

49. Shirai, "Brief," 217; Haga, 247; Nihon Shokubutsu Gakkai, *Hyakka*, 7; Marcon, *Knowledge*, 209–210.

50. Marcon, *Knowledge*, 8, 228, 243.

51. Havens, *Nishi*, 100–103.

52. Bleichmar, *Visible*, 9. See also Bleichmar, "Conquistadors," 250–253.

53. Tachibana, *Kusubana*.

54. Iwasaki, *Honzō zufu*; Iwasaki, *Bukō*.

55. Fukuoka, 2–3, 6; Marcon, *Knowledge*, 200.

56. Siebold, *Florilegium*, 1:vii, 12. See Rix, 85.

57. Takatori, 1:30.

58. Williams, 4; Scharf, 86–87.

59. Hirohito, "Linné," 115; Scharf, 92; Williams, 7. If there were moral qualms about the idea of plant sex, Linnaeus' theories helped calm them. As Williams points out, Vaillant used the pistachio tree to illustrate plant sex. Williams, 12–13.

60. Williams, 28. See 24; Muntschick, 73; Scharf, 86, 92.

61. Scharf, 102–104. Jean-Baptiste Lamarck (1744–1829), in *Flore françoise*, 3 vols. (Paris: L'imprimere royale, 1778), combined Linnaeus' artificial approach with a natural system to describe four thousand plants. Written in French, this work sold twelve thousand copies and required reprinting in 1795. Scharf, 95–97; Williams, 61–62.

62. See Kihara, 235; Itoigawa, 2.

63. Nelson, 53. See Thunberg, *Japan*.

64. Thunberg, *Flora*. See Kowenhoven and Forrer, 76; Kodansha, 5:313; Levy-Yamamori and Taaffe, 32; Kimura Yōjirō, 13–16. See Watanabe, 84, on Thunberg's misidentification of Japanese knotweed.

65. Nihon Shokubutsu Gakkai, *Hyakka*, 3; Spongberg, 100; Kimura and Kitamura, 16; Crane, 201. Stearn, 111–112, points out the ways in which Thunberg made extensive use of the writings of Engelbert Kaempfer.

66. Muntschick, 76; Marcon, *Knowledge*, 135, 139; Ueno Masuzō, *Jidai*, 157.

67. Bowers, 90. See Spongberg, 99–100; Tsukamoto and Creech, 134. Shortly after Thunberg left Japan, Isaac Titsingh (1745–1812) served as Dutch trade commissioner at Nagasaki for three different periods between 1779 and 1784. Friendly with Japanese scholars of Western learning, Titsingh wrote two books in French on everyday life in Japan. Kodansha, 8:30.

68. Hirohito, "Linné," 117. Among Siebold's most prominent students was Itō Genboku

(1800–1871), a leading expert on Western medicine in the mid-nineteenth century who became physician to the shogun in 1858 and translated medical works from Dutch into Japanese. Kodansha, 3:352.

69. Spongberg, 102.

70. Watanabe, 75; Bartholomew, 31; *Kodansha*, 3:26, 7:192–193. For Takahashi on the exclusion of foreigners, see Wakabayashi, 101–107.

71. Tsukamoto and Creech, 135.

72. Kowenhoven and Forrer, 10.

73. Tsukamoto and Creech, 135.

74. Kimura and Kitamura, 1–2, 16; Plutschow, 15; Kodansha, 1:76.

75. Kimura and Kitamura, 6, 17; Spongberg, 100; Plutschow, 15.

76. Marcon, *Knowledge*, 263. See Rietbergen, 220.

77. Kowenhoven and Forrer, 68–72; Tsukamoto and Creech, 136; Spongberg, 102–103; Ishiyama, 208–209.

78. Barnes, 118.

79. Ibid. Levy-Yamamori and Taaffe, 32. Siebold's name is also carried by species of abalone and water snake.

80. *Kodansha*, 1:120.

81. Itō Ihee, xiii; Marcon, *Knowledge*, 179.

82. Tsukamoto and Creech, 22.

83. Nihon Shokubutsu Gakkai, *Hyakka*, 20. See Ohnuki-Tierney on how flowers, especially roses, can symbolize dictatorial leaders.

84. Kihara, 164, 235–236; Itō Ihee, xii–xvii. See Tsukamoto and Creech, 131.

85. Nihon Shokubutsu Gakkai, *Hyakka*, 21.

86. Marcon, *Knowledge*, 163, 165; Nihon Shokubutsu Gakkai, *Hyakka*, 21.

87. Nihon Shokubutsu Gakkai, *Hyakka*, 21.

88. Tsukamoto and Creech, 112–113.

89. Ibid., 82, 85, 125, 145; Nelson, 61.

90. Kihara, 164; Bartlett and Shohara, 63.

91. Udagawa Yōan, *Botanika kyō*, manuscript, quoted in Ueno Masuzō, *Jidai*, 9. Udagawa probably translated the Dutch *plantkunde* as botany. See Miller, 25.

92. Ueno Masuzō, *Jidai*, 125–126. See 9–10; Marcon, *Knowledge*, 257–258.

93. Nihon Shokubutsu Gakkai, *Hyakka*, 7; Kimura and Kitamura, 17–18.

94. Makino, "Preface," 1:15.

95. Kihara, 165; Nihon Shokubutsu Gakkai, *Hyakka*, 7; Bartlett and Shohara, 65; Ueno Masuzō, *Nihon*, 134–135.

96. Shirahata, 270; Bartholomew, 58; Iwatsuki Kunio, *Nihon*, 106; Kimura and Kitamura, 17–19; Marcon, *Knowledge*, 266.

97. Ueno Masuzō, *Nihon*, 137. See Ōba, *Koishikawa*, 96–109.

98. Kitamura Masami, quoted in Reitan.

99. See Marcon, *Knowledge*, 275, 302.

100. Scharf, 75.

Chapter 3: Seeking Japanese Plants in Europe and North America

1. Enrique Dussel, quoted in Maldanado Torres, 210. See Watanabe, 98; Bleichmar, *Visible Empire*, 9, 189; Mueggler, 18–20.

2. Spongberg, 84–85.

3. Nelson, 58.

4. Cook, 318; Mueggler, 18–20; Spongberg, 91–96.

5. Williams, 1–3.

6. *List of Plants,* A2.

7. Ibid., 11–24.

8. Tsukamoto and Creech, 105, 110–111.

9. Creech, "Commentary," xi; Nelson, 59–63.

10. Spongberg, 110; Mueggler, 19.

11. Watanabe, 51; Kowenhoven and Forrer, 48, 55, 87.

12. Elliott, 20.

13. Anderson, 110; Kimura and Kitamura, 35; Watanabe, 51, 99; Nelson, 53; Kowenhoven and Forrer, 80; Wilson, *Lilies,* 6. Spongberg, 104–107, provides details of other plant introductions by Siebold.

14. See Mueggler, 32. For Hooker, see Endersby.

15. Alexander, 24.

16. Spongberg, 156–157.

17. John Gould Veitch, 487–498.

18. Barnes, 123–125; Tsukamoto and Creech, 145.

19. Nelson, 57–58, 63.

20. Spongberg, 158. See Nelson, 60.

21. Fortune, 387. See Spongberg, 122, 125; Barnes, 119; Tsukamoto and Creech, 141; Herries, 2.

22. Robert Fortune (1862), quoted in Coats, *Quest,* 74. See Elliott, 20.

23. Barnes, 119; Tsukamoto and Creech, 140; Nelson, 65.

24. Robert Fortune, quoted in Nihon Shokubutsu Gakkai, *Hyakka,* 20 (retranslated from Japanese).

25. Hooker, 327–350. See John Gould Veitch, 489.

26. Oliver, 163–170.

27. Nelson, 54; Coats, *Quest,* 79; Barnes, 125; Charlotte Rowley, "Richard Oldham: The Last Botanical Collector," Royal Botanic Gardens, Kew, accessed August 28, 2019, http://apps.kew.org.

28. Barnes, 121–123; Boufford, "Japanese," 30; Coats, *Quest,* 75.

29. Ueno, *Nihon,* 139.

30. Ōi, 7; Barnes, 121–123; Herries, 2; Nelson, 54.

31. Barnes, 126.

32. See Mitford, *Mitford's Japan;* Mitford, *Tales;* Mitford, *Bamboo.*

33. Ueno, *Nihon,* 140; Iwatsuki et al., 1:ix. See Barnes, 126–127.

34. Ueno, *Nihon,* 140. Franchet published other volumes based on plants collected by French missionaries and other travelers in the second half of the nineteenth century. See Spongberg, 177.

35. Savatier, 232.

36. Ōi, 8; Nihon Shokubutsu Gakkai, *Hyakunen,* 107; Whittle, 157; Boufford, "Japanese," 30.

37. Spongberg, 172–173; Barnes, 128–129; Alexander, 24.

38. Del Tredici, "Larz," 3.

39. Alexander, 25; Abe.

40. Nelson, 62. See 52, 56, 61; Alexander, 25.

41. Creech, "Foreword," 6.

42. Del Tredici, "Introduction."

43. Butler, "Garden," 265.

44. Spongberg, 71. The botanical garden in Cambridge was converted to married student housing after World War Two.

45. Morrow to his parents, March 3, 1853, in Cole, 255.

46. Morrow diary, April 3, 1854, in Cole, 155.

47. Tsukamoto and Creech, 137–138. See Cole, 159, 214, 227.

48. Creech, *Ornamental,* 4; Del Tredici, "Introduction"; Tsukamoto and Creech, 142–143; Spongberg, 145.

49. Barnes, 121; Spongberg, 145–146, 150; Del Tredici, "Introduction." Spongberg, 146, notes that the umbrella pine is not a pine but a member of the *Sciadopitys* genus. Fossils from Germany show that this species grew 30 million years before the present.

50. Sargent, "The Forest: Some Japanese," 537–538, quoted in Butler, "Garden," 257.

51. Butler, "Plant," 113, 144; Butler, "Garden," 287.

52. Spongberg, 153.

53. Del Tredici, "Introduction"; Spongberg, 161; Tsukamoto and Creech, 145.

54. Del Tredici, "Introduction."

55. Tsukamoto and Creech, 146; Barnes, 125; Del Tredici, "Introduction."

56. Spongberg, 162; Del Tredici, "Introduction."

57. Thornton, 4–10.

58. Levy-Yamamori and Taaffe, 19; Cunningham, "Exploration," 228; "Nancy Reagan Will Present a Cherry Tree to Japanese," posted February 23, 1981, United Press International archives, accessed August 28, 2019, https://www.upi.com.

59. Takahashi and Kawahara, 10, 14.

60. Ibid., 14; Spongberg, 168, 176.

61. Elliott, 20; Del Tredici, "Larz," 3.

62. Kramer to Hooker, December 31, 1870. "Letter from Carl Kramer to Sir Joseph Dalton Hooker," Royal Botanic Gardens, Kew, Archives: Directors' Correspondence, accessed August 28, 2019, https://plants.jstor.org.

63. *London and China Telegraph,* February 26, 1872, 167.

64. "Kramer, Carl," Meiji Portraits, accessed August 28, 2019, www.meiji-portraits.de.

65. Boehmer, 304.

66. Tsukamoto and Creech, 146–149, 165; Creech, "Pioneer," 381–383; Elliott, 21.

67. Creech, "Pioneer," 380–381.

68. L. Boehmer and Company, 9.

69. Ibid.

70. Creech, "Pioneer," 383.

71. Elliott, 21; Tsukamoto and Creech, 149; Del Tredici, "Larz," 3; Spongberg, 227; Creech, "Pioneer," 383; "Bulbs and Flowers," Yokohama Nursery Company, accessed August 28, 2019, www.yokohamaueki.co.jp.

72. Yokohama, 1901–1924, 1936.

73. Warner, ix–x; "Celebrating," 1, 4.

74. Georgeson, 331.

75. Yamanaka and Company, 4.

76. James H. Veitch, 122.

77. Ibid., 107.

78. Ibid., 115.

79. Tsukamoto and Creech, 130, 146, 170; Spongberg, 171–176.

80. Sargent, *Forest Flora,* 2–3. Sargent's records occupy eight linear feet in the archives of the Arnold Arboretum Horticultural Library in Jamaica Plain, Boston.

81. Creech, *Ornamental,* 13–18; Tsukamoto and Creech, 151, 156.

82. Tsukamoto and Creech, 153–155. Ibid., 164, explains that the U.S. depends on other countries for crop diversity because only blueberry, pecan, sunflower, and cranberry are native.

83. Fairchild, 17.

84. Pearson, "Jack," 8; Spongberg, 198; Tsukamoto and Creech, 157–158.

85. Del Tredici, "Larz," 4–7, 11; "Records of the Larz Anderson Bonsai Collection, 1904–: Guide," Arnold Arboretum of Harvard University, accessed August 28, 2019, https://arboretum.harvard.edu. Larz and Isabel Anderson acquired a large collection of Japanese art objects during their 1897 honeymoon, one of four trips they made to Japan. Much of the collection is contained in Anderson House along Massachusetts Avenue on Embassy Row in Washington, D.C. "'Wonderland of the World': The Andersons and Japan," accessed February 9, 2019, https://networks.h-net.org.

86. Spongberg, 218.

87. Sutton, 253.

88. Wilson, *Cherries,* vii–ix.

89. Pearson, *Arnold,* 8; Creech, *Ornamental,* 4–5; Sutton, 257.

90. Cunningham, *Frank,* 202–203.

91. Tsukamoto and Creech, 161–162.

92. Ibid., 114. See Healey.

93. Creech, *Southern, Plant Explorations,* 1–4, 20–43. See Spongberg, 237; Creech, *Ornamental,* 22, 43–64.

94. Ōfuna, 1–5.

95. Tsukamoto and Creech, 119, 165; Spongberg, 237–238.

96. Cunningham, "Exploration," 28; *Washington Post* online, March 20, 2012.

Chapter 4: Foundations of Plant Biology in Modern Japan

1. Tanaka Yoshio, quoted in Ueno Masuzō, *Nihon,* 146. See Ueno Masuzō, *Jidai,* 132–134; Kihara, 253; Bartholomew, 129.

2. Tanaka Yoshio. Tanaka focused on foodstuffs, not economic plants (materials and supplies).

3. Ueno Masuzō, *Nihon,* 147; Ōba, *Koishikawa,* 45.

4. Ueno Masuzō, *Nihon,* 142.

5. Itoigawa, 4.

6. Kokuritsu Kagaku Hakubutsukan, *Hyakunenshi,* 638–639; Nihon Shokubutsu Gakkai, *Hyakka,* 8.

7. Havens, *Nishi,* 35, 227.

8. Nagakubo, 206.

9. Nihon Shokubutsu Gakkai, *Hyakka,* 8; Ueno Masuzō, *Jidai,* 137–138; Nihon Shokubutsu Gakkai, *Hyakunen,* 52, 185; Kokuritsu Kagaku Hakubutsukan, *Hyakunenshi,* 34; Tawara, 37; Iwatsuki Kunio, *Nihon,* 105.

10. Nagakubo, 170; Kokuritsu Kagaku Hakubutsukan, *Hyakunenshi,* 640.

11. Kokuritsu Kagaku Hakubutsukan, *Hyakunenshi,* 639; Nihon Shokubutsu Gakkai, *Hyakunen,* 52; Ueno Masuzō, *Jidai,* 173–174. Ōkubo Samurō left Tokyo Imperial University in 1895 to become a teacher at Tokyo Higher Normal School and gave up his career as a botanist. Ōba, *Koishikawa,* 96.

12. Tawara, 38.

13. Nihon Shokubutsu Gakkai, *Hyakka,* 8, 5–8; Nihon Shokubutsu Gakkai, *Hyakunen,* 5–8; Ueno Masuzō, *Nihon,* 151; Ōi, 8.

14. Imahori, 1.

15. Nihon Shokubutsu Gakkai, *Hyakunen,* 12.

16. Ibid., iii.

17. Itō Tokutarō (1888), quoted in Ueno Masuzō, *Nihon,* 150.

18. Yatabe, "Few," 355.

19. Yatabe Ryōkichi, quoted in Nihon Shokubutsu Gakkai, *Hyakunen,* 50.

20. Ōba, *Koishikawa,* 90–93; Nagakubo, 3, 202–205.

21. Nagakubo, 3.

22. Matsumura Jinzō, quoted in Kim, 73. See Masuda.

23. Ōba, *Koishikawa,* 90; Nagakubo, 2.

24. Ōba, *Koishikawa,* 93; Nagakubo, 205–209; Nihon Shokubutsu Gakkai, *Hyakunen,* 55, 81.

25. Ueno Masuzō, *Nihon,* 138; Ōba, *Shokubutsu bunka,* 269–270; Nihon Shokubutsu Gakkai, *Hyakunen,* 54–59.

26. Nihon Shokubutsu Gakkai, *Hyakunen,* 53, 63; Kokuritsu Kagaku Hakubutsukan, *Hyakunenshi,* 640.

27. Iwatsuki Kunio, *Nihon,* 124–126.

28. Miyabe and Kudō; Hara, "Preliminary," 685; Toyokuni, 127.

29. Nihon Shokubutsu Gakkai, *Hyakunen,* 62–64.

30. Kōchi Shinbunsha, I, 10, 187–188.

31. Makino Tomitarō, quoted ibid., 69.

32. Nagakubo, 160–161; Tawara, 46; Nihon Shokubutsu Gakkai, *Hyakka,* 30.

33. Kōchi Shinbunsha, 72–74; Tawara, 39–40; Nagakubo, 161–162.

34. Kōchi Shinbunsha, 76.

35. Matsumura Jinzō, quoted ibid., 75.

36. Yatabe Ryōkichi, quoted in Nagakubo, 162.

37. Kōchi Shinbunsha, 74–85; Tawara, 34–38; Nagakubo, 162–163.

38. Nagakubo, 164–166. See 158, 174; Kōchi Shinbunsha, 86; Tawara, 32, 47–48. Makino's income from his initial appointment at the University of Tokyo in 1893 was insufficient to support his wife and growing family, which eventually included thirteen children.

39. Makino Tomitarō, quoted in Nagakubo, 172.

40. Nagakubo, 234.

41. Ibid., 171.

42. Nihon Shokubutsu Gakkai, *Hyakunen,* 110–112; Ōba, *Shokubutsu bunka,* 433–434; Crane, 67. Fujii Kenjirō later founded the journal *Cytologia* and rose to full professor at the University of Tokyo.

43. Nihon Shokubutsu Gakkai, *Hyakka,* 5, 9; Nihon Shokubutsu Gakkai, *Hyakunen,* 81; Tamura, 17–24; Nagakubo, 175.

44. Miyoshi, 409–410. See Nihon Shokubutsu Gakkai, *Hyakunen,* 111; Ueno Masuzō, *Nihon,* 150.

45. Ōba, *Shokubutsu bunka,* 424.

46. Ueno Masuzō, *Nihon,* 150.

47. Nihon Shokubutsu Gakkai, *Hyakka,* 9.

48. Métailié, 211.

49. Thornber, *Empire,* 34–42.

50. Nihon Shokubutsu Gakkai, *Hyakunen,* 73–74; Itoigawa, 12.

51. Iwatsuki Kunio, *Nihon,* 136.

52. Ibid., 11.

53. Kokuritsu Kagaku Hakubutsukan, *Hyakunenshi,* 49. See 46.

54. Ōba, *Shokubutsu bunka,* 55–56; Iwatsuki, 105–106. The College of Science became the Faculty of Science in 1919.

55. Ōba, *Shokubutsu bunka,* 43. See Kawakami, 7.

56. Miyake Kiichi, 1477; Ōba, *Shokubutsu bunka,* 46–48; Kawakami, 7; Iwatsuki Kunio, *Nihon,* 106; Nihon Shokubutsu Gakkai, *Hyakunen,* 13.

57. Ōba, *Shokubutsu bunka,* 48.

58. Ibid., 49, 101.

59. Ōba, *Koishikawa,* 101, 107–108; Nihon Shokubutsu Gakkai, *Hyakunen,* 87–88.

60. Nihon Shokubutsu Gakkai, *Hyakunen,* 74–75; "The Riken Story," Riken, accessed August 28, 2019, http://www.riken.jp.

61. Ōba, *Shokubutsu bunka,* 209; Ueno Masuzō, *Nihon,* 149; Koyama, *Available,* 939; Nihon Shokubutsu Gakkai, *Hyakunen,* 81.

62. Ōba, *Shokubutsu bunka,* 505–507; Nihon Shokubutsu Gakkai, *Hyakunen,* 13; Nihon Shokubutsu Gakkai, *Hyakka,* 13.

63. Ōba, *Shokubutsu bunka,* 252. See Nihon Shokubutsu Gakkai, *Hyakka,* 9.

64. "Professor Tsuneyoshi Kuroiwa Elected to the Japan Academy," Rikkyo University, accessed August 28, 2019, http://english.rikkyo.ac.jp. Ōba, *Shokubutsu bunka,* 165; Nihon Shokubutsu Gakkai, *Hyakka,* 10.

65. Nihon Shokubutsu Gakkai, *Hyakka,* 10; Ōba, *Shokubutsu bunka,* 433–434.

66. Nihon Shokubutsu Gakkai, *Hyakka,* 10; Ōba, *Shokubutsu bunka,* 183–185.

67. "Masayoshi Sugiura," Nagoya University, accessed August 28, 2019, http://www.gene.nagoya-u.ac.jp.

68. Nihon Shokubutsu Gakkai, *Hyakka,* 11.

69. Nihon Shokubutsu Gakkai, *Hyakunen,* 88.

70. Ibid., 88–89; Ōba, *Shokubutsu bunka,* 527.

71. Ōba, *Shokubutsu bunka,* 25.

72. Nihon Shokubutsu Gakkai, *Hyakunen,* 107–108

73. Ibid., 100; Ōba, *Shokubutsu bunka,* 320–321.

74. Okamura Kintarō, "Kaisō mandan" (1929), quoted in Nihon Shokubutsu Gakkai, *Hyakunen,* 91.

75. Ōba, *Shokubutsu bunka,* 128–129; Nihon Shokubutsu Gakkai, *Hyakunen,* 91–93.

76. Ōba, *Shokubutsu bunka,* 103; Nihon Shokubutsu Gakkai, *Hyakunen,* 92, 95.

77. Ueno Masuzō, *Nihon,* 149, 159; Wulf, 7.

78. Wulf, 9–10; Ōba, *Shokubutsu bunka,* 217, 374.

Chapter 5: Plant Biology in Japan's Age of Empire

1. See Gordon.

2. Makino Tomitarō, *Makino Tomitarō shokubutsuki* (1948), quoted in Kōchi Shinbunsha, 199.

3. Ueno, *Nihon,* 152.

4. Ōi, 8; Tawara, 17–18, 75–79; Kōchi Shinbunsha, 151–161.

5. Nihon Shokubutsu Gakkai, *Hyakka,* 30; Nagakubo, 165–166; Kōchi Shinbunsha, 184–185; Shiraiwa.

6. Tawara, 65–66; Ueno, *Nihon,* 152–153.
7. Ōba, *Koishikawa,* 98.
8. Kōchi Shinbunsha, 146. See 145, 149, 159; Tawara, 66–67.
9. Tawara, 67; Kōchi Shinbunsha, 149, 161–162.
10. Ōba, *Shokubutsu bunka,* 481–482; Ueno, *Nihon,* 153.
11. Ueno, *Nihon,* 155.
12. Ōba, *Koishikawa,* 98; Tawara, 58–59; Nihon Shokubutsu Gakkai, *Hyakka,* 30–31; Nihon Shokubutsu Gakkai, *Hyakunen,* 81–82.
13. Tachibana Setsu, 4.
14. Ueno, 153; Nihon Shokubutsu Gakkai, *Hyakunen,* 85–86.
15. Nihon Shokubutsu Gakkai, *Hyakka,* 31; Tawara, 10, 62–63; Kamimura, 348–357.
16. Iwatsuki, 153–154; Tawara, 68.
17. Nihon Shokubutsu Gakkai, *Hyakka,* 30; Tawara, 49, 67–69; Iwatsuki, 153–154.
18. Tawara, 9; Kōchi Shinbunsha, 164.
19. Kadota Yūichi (2012), quoted in Kōchi Shinbunsha, 197.
20. Tsurumi, "Minakata-Mandala," 172; Matsui and Iwasaki, 8–9.
21. Tsurumi, "Comparative," 86, 90. For a study of Minakata's Buddhist ecology, see Godart, 92–103.
22. Minakata Kumagusu, quoted in Kokuritsu Kagaku Hakubutsukan, *Minakata,* 11.
23. Matsui and Iwasaki, 9.
24. Kokuritsu Kagaku Hakubutsukan, *Minakata,* 3–5; Tsurumi, "Comparative," 72; Matsui and Iwasaki, 16–22; Blacker, 139–143. See Nakazawa for a reinterpretation of Minakata's career.
25. Hosoya Tsuyoshi, quoted in *Japan Times* online, January 28, 2018.
26. Kokuritsu Kagaku Hakubutsukan, *Minakata,* 7–9, 18; Tsurumi, "Minakata-Mandala," 176–178. Minakata's hypothesis that slime molds were protozoa has been reinterpreted by some to mean that they were a eukaryotic life form intermediate between plants and animals. See Woese and Fox, 5088–5090.
27. Seibutsugaku Gokenkyūjo.
28. Kokuritsu Kagaku Hakubutsukan, *Minakata,* 15–21.
29. Ibid., 10, 25; Matsui and Iwasaki, 10; Tsurumi, "Comparative," 94; Fridell.
30. *Japan Times* online, January 28, 2018; Matsui and Iwasaki, 10–11; Watari-um, 4.
31. See Takamatsu.
32. Hayata, *Vegetation,* 110.
33. Hayata, *Botany,* 5, 28. See Lee, "Provincialising," 436–445.
34. Nihon Shokubutsu Gakkai, *Hyakunen,* 82; Ōba, *Koishikawa,* 102; Ōba, *Shokubutsu bunka,* 398–399, 479–481.
35. Nihon Shokubutsu Gakkai, *Hyakunen,* 82–83; Ōba, *Shokubutsu bunka,* 453–455.
36. Nihon Shokubutsu Gakkai, *Hyakunen,* 108; Ōba, *Shokubutsu bunka,* 11–12.
37. Miyachi, 1.
38. Nihon Shokubutsu Gakkai, *Hyakunen,* 112–115; Ōba, *Shokubutsu bunka,* 133–134.
39. Nihon Shokubutsu Gakkai, *Hyakunen,* 155–156.
40. Ōba, *Shokubutsu bunka,* 328, 470; Nihon Shokubutsu Gakkai, *Hyakunen,* 156–157.
41. Nihon Shokubutsu Gakkai, *Hyakunen,* 192–194, 197; Ōba, *Shokubutsu bunka,* 374.
42. Nihon Shokubutsu Gakkai, *Hyakunen,* 195–199, 208.
43. Schiebinger and Swan, 3. See Nihon Shokubutsu Gakkai, *Hyakka,* 22.
44. Schiebinger, 7, 11. A classic study of colonial botany is Brockway.

45. Bleichmar, *Visible Empire,* 5, 189.

46. Parsons, "Wildness," 657. Mizuno Hiromi offers an alternative perspective on Japanese science policy in the 1920s to 1940s, emphasizing the interplay of science, imperialism, and nationalism. See Mizuno.

47. A half century ago George Basalla's concept of "colonial science" as a stage in scientific and technological transfer from metropole to colony was influential, but it has long since outlived its usefulness because of its culture-centrism, its emphasis on the metropole as unchanging, and its failure to recognize the diversity of science and technology in colonized societies. See Basalla, 611–622.

48. Ōi, 8; Nihon Shokubutsu Gakkai, *Hyakunen,* 15, 19; Ōba, *Shokubutsu bunka,* 162; Iwatsuki et al., 1:ix. *The Botanical Magazine* was a key forum for research on Japan's empire until wartime stringencies forced it to suspend publication in June 1944. Its publisher, Sanshūsha, was destroyed in an air raid in November 1944; the journal resumed in March 1946 with Hokuryūkan as publisher. See Nihon Shokubutsu Gakkai, *Hyakunen,* 26–27.

49. Ōba, *Shokubutsu bunka,* 162; Nihon Shokubutsu Gakkai, *Hyakunen,* 82.

50. Honda, 1–4. Hagino Toshio (1920–) had a similarly Pan-Asian perspective on forestry, in *Chōsen Manshū Taiwan ringyō hattatsushiron* (1964).

51. Horikawa, *Distributional,* 2–4. The *Dai Kan Wa jiten* is the standard dictionary of premodern Chinese and is considered a monument of painstaking, reliable scholarship.

52. Ivings, 3.

53. Sugawara, *Plants,* 5–41.

54. Samukawa Kōtarō, quoted in Weiner, 258. See Ōba, *Shokubutsu bunka,* 276, 375; Ōba, *Koishikawa,* 97. Sugawara published a condensed version of his four-volume *Karafuto shokubutsu zushi* in English and Japanese as *Plants of Saghalien = Karafuto no shokubutsu* in 1937.

55. Nihon Shokubutsu Gakkai, *Hyakunen,* 93; Ōba, *Shokubutsu bunka,* 350–351.

56. See Saraiva on crops and flocks in European colonies in Africa and Nazi-controlled territories in Poland and the Soviet Union.

57. Henry, 14–118.

58. Ōba, *Koishikawa,* 97; Ōba, *Shokubutsu bunka,* 411–412.

59. Hayata, *Flora,* 4–5.

60. Ōhashi, "Bunzō Hayata," 1–27; Ōba, *Koishikawa,* 99.

61. Ōba, *Koishikawa,* 99–101; Nihon Shokubutsu Gakkai, *Hyakunen,* 81.

62. Ōba, *Shokubutsu bunka,* 467, 547; Nihon Shokubutsu Gakkai, *Hyakunen,* 65.

63. Masamune, 8–11.

64. Hu Wenqing, 190; Murasaki, 24.

65. See Uchida.

66. Nakai.

67. Hara, "Takenoshin Nakai," 1–3; Ōi, 8; Ōba, *Koishikawa,* 101–102.

68. Hotta Teikichi, 58–59; Ōba, *Shokubutsu bunka,* 96. 447.

69. Lee, "Mutual," 179. See Ishidoya, 457–490.

70. See Havens, *Parkscapes,* 84. On tourist accounts of Korea, see McDonald.

71. Ōba, *Koishikawa,* 102.

72. Ibid., 106.

73. Hara, "Takenoshin Nakai," 2; Ōba, *Shokubutsu bunka,* 364–366.

74. "Unmarked."

75. Nihon Shokubutsu Gakkai, *Hyakunen,* 84.

76. Ibid., 82–84; Ōba, *Koishikawa,* 102; Ōba, *Shokubutsu bunka,* 179.

77. John Young. See Louise Young.
78. Nihon Shokubutsu Gakkai, *Hyakunen*, 84–85.

Chapter 6: Plant Biologists in an Era of Specialization

1. Azuma Yoshirō interview, December 17, 2017.
2. Ibid.
3. Iwatsuki Kunio, 30.
4. Ōba, *Shokubutsu bunka*, 280–281.
5. Ibid., 11–12; see Hayami and Kantha, 90.
6. Garfield, "Citation," 10:282–292.
7. Hayami and Kantha, 84; Chaffey; "Want a Nobel Prize?"
8. Nihon Shokubutsu Gakkai, *Hyakunen*, 68–73.
9. Ibid., 56–58, 105–106.
10. Ibid., 59–60.
11. Ōba, *Shokubutsu bunka*, 554.
12. Nihon Shokubutsu Gakkai, *Hyakunen*, 61–62.
13. Ōba, *Shokubutsu bunka*, 540.
14. "Sugadaira Montane Research Center, University of Tsukuba," University of Tsukuba, accessed August 28, 2019, http://www.sugadaira.tsukuba.ac.jp. See Nihon Shokubutsu Gakkai, *Hyakunen*, 65–68.
15. Nihon Shokubutsu Gakkai, *Hyakunen*, 207; Ōba, *Koishikawa*, 106–107; Iwatsuki et al., 1:ix; Tatsumi.
16. "Kinki Nihon tsūrisuto," Kinki Nihon Tetsudō, accessed August 28, 2019, http://www.knt.co.jp ; "Nihon Kōseigaku Kyōkai," accessed September 2, 2018, Nihon Kōseigaku Kyōkai, http://photon.umin.jp ; Nihon Shokubutsu Gakkai, *Hyakunen*, 30–35; Miyachi, 1.
17. Japan Association of Botanical Gardens, 3, lists sixty-three members, including nurseries.
18. Nihon Shokubutsu Gakkai, *Hyakka*, 23.
19. Iwatsuki Kunio, 87–88, 99–100.
20. Ibid., 30. See 11; Nihon Shokubutsu Gakkai, *Hyakka*, 23.
21. Nihon Shokubutsu Gakkai, *Hyakka*, 34–35, 107–111, 122–123.
22. "Herbarium of the University of Tokyo," University of Tokyo, accessed August 28, 2019, http://umdb.um.u-tokyo.ac.jp. See Tokyo Daigaku, *Type Specimens;* Tokyo Daigaku, *Fossil Plants;* Ōhashi, *List*, 1–26.
23. Yumenoshima, 1–2; *Yomiuri shinbun* online, March 17, 2016.
24. *Itabashi Akatsuka*.
25. "Herbarium of the University of Tokyo," University of Tokyo, accessed August 28, 2019, http://umdb.um.u-tokyo.ac.jp. See Iwatsuki Kunio, 130.
26. Tokyo Metropolitan, viii; *Jindai*, 1–16; "Jindai Botanical Gardens," Tokyo Metropolitan Government, accessed August 28, 2019, http://www.kensetsu.metro.tokyo.jp.
27. Tama, 1–2.
28. "Research Institute," Tokyo University of Agriculture, accessed August 28, 2019, https://www.nodai.ac.jp.
29. Iwatsuki Kunio, 130.
30. Kokuritsu Kagaku Hakubutsukan, *Tsukuba*, 1–2.
31. "Botanical Garden," Tohoku University, accessed August 28, 2019, https://www.tohoku.ac.jp. See Iwatsuki Kunio, 126–127.

32. "Higashiyama Zoo and Botanical Gardens," *Aichi Now,* accessed August 28, 2019, https://www.aichi-now.jp. See Iwatsuki Kunio, 145–146.

33. "Kyoto Botanical Garden," Wikipedia, accessed August 28, 2019, https://en.wikipedia.org. See Iwatsuki Kunio, 146–147.

34. "Botanical Gardens," Osaka City University, accessed August 28, 2019, https://www.osaka-cu.ac.jp.

35. Kobe Municipal, 1–31; "Kobe Travel Facilities," *Feel Kobe,* accessed August 28, 2019, http://plus.feel-kobe.jp.

36. "Hiroshima Botanical Garden," *Explore Hiroshima,* accessed August 28, 2019, https://www.hiroshima-navi.or.jp.

37. Arima, 590.

38. Ibid., 590–591.

39. Hirano, 582–583.

40. *Yomiuri shinbun* online, May 30, 2017.

41. Ibid.; *Mainichi shinbun* online, May 6, 2019.

42. "Japan Boosts"; Phillips.

43. "Japanese Researchers."

44. Ōba, *Shokubutsu bunka,* 455–457.

45. Nihon Shokubutsu Gakkai, *Hyakunen,* 116–117.

46. Ibid., 121.

47. Ōba, *Shokubutsu bunka,* 165, 235–236.

48. Nihon Shokubutsu Gakkai, *Hyakunen,* 118–121.

49. "Kutsukake Kazuhiro," Research Map Japan, accessed August 28, 2019, https://researchmap.jp ; Martin Nowak has argued that cooperation stands side by side with mutation and selection as a basic component of evolution. Nowak, "The Mathematics of Cooperation," lecture, Harvard University, February 11, 2019.

50. Nihon Shokubutsu Gakkai, *Hyakunen,* 117, 120, 154–155.

51. "Outline and History," National Institute of Genetics, accessed August 28, 2019, https://www.nig.ac.jp.

52. Nagai et al., 1.

53. Ibid., 10. See 4, 7, 13–14, 17.

54. "Outline and History," National Institute of Genetics, accessed August 28, 2019, https://www.nig.ac.jp.

55. Hirohito, "Linné," 120.

56. Kokuritsu Kagaku Hakubutsukan, *Tsukuba,* 30; Nihon Shokubutsu Gakkai, *Hyakka,* iii–iv, 9; Iwatsuki Kunio, 33–35. Iwatsuki Kunio, 87, points out that the Ministry of Agriculture, Forestry, and Fisheries maintains a gene bank to preserve plants systematically. The Ministry of Health, Labor, and Welfare preserves medicinal plants. Botanical gardens often struggle to preserve species verging on extinction.

57. Nihon Shokubutsu Bunrui, 27, 34–37.

58. Nishikawa, iii Nihon Shokubutsu Gakkai, *Hyakunen,* 87.

59. Matsunaga, 142.

60. Ibid., 142–145, 161.

61. Satō, 265. See 266–271.

62. "Nishina Zao," Riken, accessed August 28, 2019, www.riken.jp ; "Japan Tests"; *Japan Today* online, July 7, 2014; "Ion Beam Breeding Team," Riken, accessed August 28, 2019, www.riken.jp ; Whitwam.

63. Ryall.

64. *Asahi shinbun* online, April 6, 2014.

65. *Asahi shinbun* online, October 29, 2014.

66. *Asahi shinbun* online, December 31, 2015.

67. "Biostructural Mechanism Group," Riken, accessed August 28, 2019, http://www.riken.jp. See *Princeton Alumni Weekly* 118:5 (December 6, 2017), 2.

68. Kurata, vii–x.

69. Kobayashi and Shimizu, 43. See Kurata, xi.

70. Nagao and Fukiharu, 131–132.

71. Nihon Shokubutsu Gakkai, *Hyakunen,* 90. Japan has an estimated 2,000 species of lichens. Nihon Shokubutsu Bunrui, 38.

72. *Japan Times* online, August 6, 2018.

73. *South China Morning Post* online, August 11, 2018.

74. "Kamakura and Early Algology," National Museum of Nature and Science, accessed August 28, 2019, https://www.kahaku.go.jp. Makino Tomitarō, the "Father of Japanese Botany," once sent Okamura a red rubber bladder from inside a football, asking him to identify this curious "alga." Okamura was not fooled, replying that the specimen was *Footbalia makinoi.*

75. Konishi, 113–115.

76. Hirose and Yamagishi; Nihon Shokubutsu Gakkai, *Hyakunen,* 98–99.

77. Takeda Yakuhin, 5.

78. Ichiro Arai et al., "Current Status of Kampo (Japanese Herbal) Medicines in Japanese Clinical Practice Guidelines," posted July 2009, Research Gate, accessed August 28, 2019, www.researchgate.net.

79. "Nippon Shinyaku Yamashina Shokubutsu Shiryōkan," Nippon Shinyaku Company, accessed August 28, 2019, http://www.nippon-shinyaku.co.jp. See Nippon Shinyaku, ii.

80. *Japan Times* online, September 15, 2013.

81. Ōba and Boufford, viii–x.

82. See Nihon Shokubutsu Gakkai, *Hyakunen,* 82–83, for a tabulation of general reference works in botany published in Japan from the 1950s through the 1970s.

83. Ibid., 82–83.

84. Kachi, ii, 6; Fujioka, 2, 543.

85. Itō Kōji, 98–99; Kurata and Nakaike.

86. Hara, *Revision,* 4–7.

87. Okita and Hollenberg, vii–viii, 3–6.

88. Kuginuki.

89. Nihon Shokubutsu Bunrui, 43–44; Uemura, 1, 4, 102.

90. Nihon Shokubutsu Gakkai, *Hyakka,* 1.

91. "Nihon Kagakusha Kaigi," Nihon Kagakusha Kaigi, accessed August 28, 2019, http://www.jsa.gr.jp. See Nihon Shokubutsu Gakkai, *Hyakunen,* 208.

92. Nihon Shokubutsu Gakkai, *Hyakunen,* 209, points out that an earlier journal, *Seitaigaku kenkyū* (Ecological research) was published at Sendai starting in 1935 but after the war was folded into the *Tohoku University Mount Hakkōda Botanical Laboratory Memoirs* (*Tohoku Daigaku Hakkōdasan Jikkenjo no kiyō*)

93. Ōba, *Shokubutsu bunka,* 374, 505–507, 522.

94. Ibid., 233; Nihon Shokubutsu Gakkai, *Hyakunen,* 197.

95. Nihon Shokubutsu Gakkai, *Hyakunen,* 197–199, 210.

96. Tohoku Shokubutsu; Nihon Shokubutsu Gakkai, *Hyakunen,* 200.

97. Nihon Shokubutsu Gakkai, *Hyakunen,* 200–206.

98. Suka, Okamoto, and Ushimaru, 51.

99. See Havens, *Parkscapes,* 70–73, 122, 141–149.

Afterword

1. Noam Chomsky, quoted in *Asia-Pacific Journal: Japan Focus* online, May 19, 2014.

2. Smith, 86, notes that *Anthropocene* is similar to the use of *Nonosphere* (world of thought) of paleontologist Pierre Teilhard de Chardin (1881–1955) to indicate the effect of human intelligence on the world.

3. Meyer points out that the International Commission on Stratigraphy in July 2018 adopted the Meghalayan age (4250 years ago to the present) as a subdivision of the Holocene epoch, without reaching agreement on using the term *Anthropocene.*

4. Hotta Mitsuru, 92.

5. Suzuki Sadao, 5, 47.

6. Ōba, *Nihon no zetsumetsu,* xv.

7. Iwatsuki et al., 1:xi.

8. Kankyōchō, 8:16.

9. Ōba, *Nihon no zetsumetsu,* xvi.

10. Iwatsuki et al., 2a:ix.

11. Yahara and colleagues' data reflect estimates in 2012. Yahara et al., 4.

12. Ōba, *Nihon no zetsumetsu,* xv.

13. Nakamura, 229.

14. Ibid., 230.

15. Nakamura, 234–236.

16. "Invasive Species of Japan," National Institute for Environmental Studies, accessed August 28, 2019, https://www.nies.go.jp.

17. *Asahi shinbun* online, November 8, 2014.

18. *Yomiuri shinbun* online, May 27, 2014.

19. "Theory."

20. Crane, 261.

21. Askins, x.

22. Hotta Mitsuru, 92–95. Deer seldom eat beech seedlings, but beech bark disease is widespread globally. "Beech Trees."

23. Tanaka Atsuo, 189. The figures are for 2009. See 199–200.

24. Ibid., 201.

25. Ibid., 202.

26. Miller-Rushing, vi, 50–64.

27. Zimmer; Kapoor.

28. Ogawa-Onishi, Berry, and Tanaka, 1728.

29. *Yomiuri shinbun* online, August 5, 2014.

30. Ibid., July 28, 2014.

31. *New York Times,* October 27, 2015, D3.

32. Nihon Shokubutsu Gakkai, *Hyakka,* 742–743. A positive development was the December 2017 discovery of a new species of East Asian hemlock that is naturally resistant to wooly adelgids, notoriously aggressive insects. Ulleungdo hemlocks (*Tsuga ulleungensis*) are

native to an island off the south coast of Korea and, like Chinese hemlocks, are not attacked by wooly adelgids. *National Geographic News* online, January 30, 2018.

33. *Asahi shinbun* online, August 31, 2015; *Japan Today* online, September 6, 2015.

34. *Yomiuri shinbun* online, May 30, 2014, August 5, 2014.

35. *Japan Times* online, March 20, 2018, June 24, 2018.

36. Yatabe Ryōkichi, quoted in Nihon Shokubutsu Gakkai, *Hyakunen,* 50.

Sources Cited

Abe, Naoko. *The Sakura Obsession: The Incredible Story of the Plant Hunter Who Saved Japan's Cherry Blossoms.* New York: Penguin Random House, 2019.

Ager, Thomas A. "Late Quaternary Vegetation and Climate History of the Central Bering Land Bridge from St. Michael Island, Western Alaska." *Quaternary Research* 60:1 (2003): 19–32.

Alcock, Rutherford B. *The Capital of the Tycoon: A Narrative of a Three Years' Residence in Japan.* 2 vols. London: Longman, Green, Longman, Roberts & Green, 1863.

Alexander, Rosemary. "A Natural Affinity—The Japanese and the British Connection." In *A Garden Bequest—Plants from Japan,* edited by Anne Farrar, 23–26. London: Japan Society, 2001.

Anderson, Mary Perle. "Early European Botanists in Japan." *Journal of the New York Botanical Garden* 8 (1907): 99–110.

Angier, Natalie. "Planet in Motion." *New York Times,* December 18, 2018, D1, D3.

Arima, Akito. "Underfunding of Basic Science in Japan." *Science,* new series 258:5082 (October 23, 1992): 590–591.

Askins, Robert A. *Saving the World's Deciduous Forests: Ecological Perspectives from East Asia, North America, and Europe.* New Haven, Conn.: Yale University Press, 2014.

Azuma Yoshirō. Personal communication with the author. Professor of Physics, Sophia University, Tokyo. December 17, 2017.

Barnes, Peter. "Japan's Botanical Sunrise: Plant Exploration around the Meiji Restoration." *Curtis's Botanical Magazine* 18:2 (2001): 117–131.

Bartholomew, James R. *The Formation of Science in Japan: Building a Research Tradition.* New Haven, Conn.: Yale University Press, 1989.

Bartlett, Harley Harris, and Shohara Hide. *Japanese Botany during the Period of Wood-Block Printing.* Los Angeles: Dawson's Book Shop, 1961.

Basalla, George. "The Spread of Western Science." *Science* 156 (1967): 611–622.

Batchelor, John, and Miyabe Kingo. "Ainu Economic Plants." *Transactions of the Asiatic Society of Japan* 21 (1893): 190–240.

Batsaki, Yota, Sarah Burke Cahalan, and Anatole Tchikine, eds. *The Botany of Empire in the Long Eighteenth Century.* Washington, D.C.: Dumbarton Oaks Research Library and Collection, 2017.

"Beech Trees Are Booming in New England: Here's Why That's a Bad Thing." *Boston.com* online, February 25, 2018.

Bicentenary Celebration of C. P. Thunberg's Visit to Japan. Tokyo: Royal Swedish Embassy and Botanical Society of Japan, 1977.

Blacker, Carmen. "Minakata Kumagusu, A Neglected Genius." *Folklore* 94:2 (1983): 139–152.

Bleichmar, Daniela. "Botanical Conquistadors." In *Worlds of Natural History,* edited by H. A. Curry, N. Jardine, J. A. Secord, and E. C. Spary, 236–254. Cambridge, Eng.: Cambridge University Press, 2018.

———. *Visible Empire: Botanical Expeditions and Visual Culture in the Hispanic Enlightenment.* Chicago: University of Chicago Press, 2012.

Bodart-Bailey, Beatrice, and Derek Massarella, eds. *The Furthest Goal: Engelbert Kaempfer's Encounter with Tokugawa Japan.* Richmond, Eng.: Curzon Press, 1995.

Boehmer, Louis. *Report of a Botanical Journey in Yesso, in 1874.* Tokyo: Kaitakushi, 1875.

Boomgaard, Peter, ed. *Empire and Science in the Making: Dutch Colonial Scholarship in Comparative Global Perspective, 1760–1830.* New York: Palgrave Macmillan, 2013.

Boufford, David E. "Eastern Asian-North American Plant Disjunctions: Opportunities for Further Investigations." *Korean Journal of Plant Taxonomy* 28:1 (1998): 49–61.

———. "Japanese Collections, Asa Gray, and the Harvard University Herbaria." In *Kuraku hakase to Sapporo no shokubutsu,* edited by Takahashi Hideki and Kawahara Noriko, 23–35. Sapporo: Hokkaido Daigaku Sōgō Hakubutsukan, 2012.

Boufford, David E., and Hideaki Ohba, eds. *Sino-Japanese Flora: Its Characteristics and Diversification.* Tokyo: University of Tokyo, 1998.

Boufford, David E., and Stephen A. Spongberg. "Eastern Asian-Eastern North American Phytogeographical Relationships—A History from the Time of Linnaeus to the Twentieth Century." *Annals of the Missouri Botanical Garden* 70 (1983): 423–439.

Bowers, John Z. *Western Medical Pioneers in Feudal Japan.* Baltimore: Johns Hopkins University Press, 1970.

Brannen, Peter. "When Life on Earth Nearly Vanished." *New York Times,* July 30, 2017, SR2.

Brockway, Lucile. *Science and Colonial Expansion: The Role of the British Royal Botanic Gardens.* New York: Academic Press, 2002.

Brombert, Victor. *In Praise of Antiheroes.* Chicago: University of Chicago Press, 1999.

Brown, Kendall H. "Territories of Play: A Short History of Japanese-Style Gardens in North America." In *Japanese-Style Gardens of the Pacific West Coast,* edited by Melba Levick and Kendall H. Brown, 8–29. New York: Rizzoli, 1999.

Butler, Sara A. "The Garden Network: George Rogers Hall's Horticultural Activism." In *Foreign Trends in American Gardens, a History of Exchange, Adaptation, and Reception,* edited by Raffaella Fabiani Giannetto, 257–287. Charlottesville: University of Virginia Press, 2016.

———. "A Plant Hunter's Legacy: Japanese Trees in a New England Landscape, 1870–1930." *Winterthur Portfolio, A Journal of American Material Culture* 50:2/3 (2016): 99–149.

"Celebrating a Century of Botany at Connecticut College." *Friends of the Connecticut College Library Newsletter* (spring 2018): 1, 4.

Chaffey, Nigel. "Noble Plant Wins Nobel Prize!" *Botany One* online, January 19, 2016.

Chaney, Ralph W. "Introduction." In *Tertiary Floras of Japan: Miocene Floras,* by Chishitsu Chōsajo, 3–6. Tokyo: Chishitsu Chōsajo, 1963.

Chishitsu Chōsajo, *Tertiary Floras of Japan: Miocene Floras.* Tokyo: Chishitsu Chōsajo, 1963.

Chung, Daisy, and Ryan T. Williams. "Talking Trees." *National Geographic* (June 2018): 26–27.

Chwastyk, Matthew W. "Future Earth." *National Geographic* (June 2018): 30–31.

Clausen, Curtis Paul. *The Parasites of* Popillia japonica *in Japan and Chosen (Korea), and Their Introduction into the United States.* Washington, D.C.: U.S. Government Printing Office, 1927.

Coats, Alice M. *Garden Shrubs and Their Histories.* New York: Dutton, 1965.

———. *The Quest for Plants: A History of the Horticultural Explorers.* London: Studio Vista, 1969. Reissued as *The Plant Hunters; Being a History of the Horticultural Pioneers, Their Quests, and Their Discoveries from the Renaissance to the Twentieth Century.* New York: McGraw-Hill, 1970.

Cockburn, Andrew. "Weed Whackers: Monsanto, Glyphosate, and the War on Invasive Species." *Harper's Magazine,* September 2015, 57–63.

Cole, Allan B., ed. *A Scientist with Perry in Japan: The Journal of Dr. James Morrow.* Chapel Hill: University of North Carolina Press, 1947.

Cook, Harold J. *Matters of Exchange: Commerce, Medicine, and Science in the Dutch Golden Age.* New Haven, Conn.: Yale University Press, 2007.

Crane, Peter R. *Ginkgo: The Tree That Time Forgot.* New Haven, Conn.: Yale University Press, 2013.

Crawford, Gary W. *Paleoethnobotany of the Kameda Peninsula Jomon.* Ann Arbor: University of Michigan Museum of Anthropology, 1983.

Creech, John L. "Commentary." In *A Brocade Pillow: Azaleas of Old Japan,* by Itō Ihee Sannojō, v–xiii. Translated by Katō Kaname. New York: Weatherhill, 1984.

———. "Foreword." In Alice M. Coats, *Garden Shrubs and Their Histories,* 6–10. New York: Dutton, 1965.

———. *Ornamental Plant Explorations, Japan, 1961.* Washington, D.C.: U.S. Department of Agriculture, 1966.

———. "Pioneer Plantsmen in Japan." *Garden, Journal of the Royal Horticultural Society* 113 (1988): 380–383.

———. *Plant Explorations: Ornamentals in Southern Japan.* Washington, D.C.: United States Department of Agriculture, 1957.

Cunningham, Isabel Shipley. *Frank N. Meyer: Plant Hunter in Asia.* Ames: Iowa State University Press, 1984.

———. "III. Exploration in Japan." *Diversity* 14 (1988): 28–29.

Curry, H. A., N. Jardine, J. A. Secord, and E. C. Spary, eds. *Worlds of Natural History.* Cambridge, Eng.: Cambridge University Press, 2018.

Darwin, Charles. *The Power of Movement in Plants.* London: John Murray, 1880; New York: New York University Press, 2010.

Davis, Jack E. *The Gulf: The Making of an American Sea.* New York: Liveright, 2017.

Del Tredici, Peter. "The Introduction of Japanese Plants into North America." *Botanical Review* 83 (2017), posted online June 8, 2017.

———. "The Larz Anderson Bonsai Collection." *Arnoldia* 49:3 (1989): 2–37.

Donoghue, Michael J., and Stephen A. Smith. "Patterns in the Assembly of Temperate Forests around the Northern Hemisphere." *Philosophical Transactions of the Royal Society B* 359:1450 (2004): 1633–1644.

Dussel, Enrique. *The Invention of the Americas: Eclipse of "The Other" and the Myth of Modernity.* Translated by Michael D. Barber. New York: Continuum, 1995.

Ebihara Atsushi. "Nihon de koyū shokubutsu ga ōi basho wa doko?" In *Uocchingu Nihon no koyū shokubutsu,* edited by Iwashina Tsukasa and Ebihara Atsushi, 140–142. Hatanoshi, Kanagawaken: Tōkai Daigaku Shuppankai, 2014.

———. "Seibutsu tayōsei hotto supotto Nihon." In *Uocchingu Nihon no koyū shokubutsu,* edited by Iwashina Tsukasa and Ebihara Atsushi, 136–139. Hatanoshi, Kanagawaken: Tōkai Daigaku Shuppankai, 2014.

Elliott, Brent. "Japanese Nurseries and Export Trade." In *A Garden Bequest—Plants from Japan,* edited by Anne Farrer, 19–22. London: Japan Society, 2001.

Endersby, Jim. *Imperial Nature: Joseph Hooker and the Practices of Victorian Science*. Chicago: University of Chicago Press, 2008.

Fairchild, David. *Three New Plant Introductions from Japan*. Washington, D.C.: U.S. Government Printing Office, 1903.

Farrer, Anne, ed. *A Garden Bequest—Plants from Japan*. London: Japan Society, 2001.

Fleming, Nic. "Which Organism Has Had the Biggest Impact on the Planet?" *BBC Earth* online, February 10, 2015.

Fridell, Wilbur M. *Shinto Shrine Mergers, 1906–12: State Shinto Moves to the Grassroots*. Tokyo: Sophia University, 1973.

Friis, Else Marie, Peter R. Crane, and Kaj Raunsgaard Pedersen. *Early Flowers and Angiosperm Evolution*. New York: Cambridge University Press, 2011.

Forseth, Irwin N., and Anne F. Innis. "Kudzu (*Pueraria montana*): History, Physiology, and Ecology Combine to Make a Major Ecosystem Threat." *Critical Reviews in Plant Sciences* 23:5 (2004): 401–413.

Fortune, Robert. *Yedo and Peking; A Narrative of the Journey to the Capitals of Japan and China*. London: John Murray, 1863.

Fujioka Shishi Hensan Iinkai. *Fujioka shishi shizen mokurokuhen*. Fujioka: Fujiokashi, 1989.

Fukui Tamotsu. "'Shobutsu ruisan' no seiritsu to denpon." *Nihon igaku zasshi* 11:4 (1965): 45–53.

Fukuoka, Maki. *The Premise of Fidelity: Science, Visuality, and Representing the Real in Nineteenth-Century Japan*. Stanford: Stanford University Press, 2012.

Gakujutsu Kenkyū Kaigi. *Scientific Japan, Past and Present*. Kyoto: Gakujutsu Kenkyū Kaigi, 1926.

Garfield, Eugene. "Citation Classics in Plant Sciences and Their Impact on Current Research" [1987]. In Eugene Garfield, *Essays of an Information Scientist: Peer Review, Refereeing, Fraud and Other Essays*, 10:282–292. Philadelphia: ISI Press, 1989.

———. *Essays of an Information Scientist: Peer Review, Refereeing, Fraud and Other Essays*. 15 vols. Philadelphia: ISI Press, 1962–1993.

Georgeson, C. C. "The Economic Plants of Japan." *American Gardener* (1891): 7–527.

Gheerbrant, Emmanuel, and Jean-Claude Rage. "Paleobiogeography of Africa: How Distinct from Gondwana and Laurasia?" *Palaeogeography, Palaeoclimatology, Palaeoecology* 241:2 (2006): 224–246.

Giannetto, Raffaella Fabiani, ed. *Foreign Trends in American Gardens: A History of Exchange, Adaptation, and Reception*. Charlottesville: University of Virginia Press, 2016.

Goble, Andrew E. *Confluences of Medicine in Medieval Japan*. Honolulu: University of Hawai'i Press, 2011.

Godart, G. Clinton. *Darwin, Dharma, and the Divine: Evolutionary Theory and Religion in Modern Japan*. Honolulu: University of Hawai'i Press, 2017.

Goldenberg, Suzanne. "Planet Earth Is Home to 8.7 Million Species, Scientists Estimate." *Guardian* online, August 23, 2011.

Goodman, Grant K. *Japan: The Dutch Experience*. Rev. ed. Dover, N.H.: Athlone Press, 1986.

Gordon, Andrew. *Labor and Imperial Democracy in Prewar Japan*. Berkeley: University of California Press, 1992.

Graham, Alan, ed. *Floristics and Paleofloristics of Asia and Eastern North America*. Amsterdam: Elsevier, 1972.

———. "Outline of the Origin and Historical Recognition of Floristic Affinities between Asia and Eastern North America." In *Floristics and Paleofloristics of Asia and Eastern North America*, edited by Alan Graham, 1–18. Amsterdam: Elsevier, 1972.

Gray, Asa. "An Account of the Botanical Specimens Collected in Japan by S. Wells Williams and James Morrow." In *Narrative of the Expedition of an American Squadron to the China Seas and Japan: Performed in the Years 1852, 1853, and 1854, under the Command of Commodore M. C. Perry, United States Navy, by Order of the Government of the United States,* by Matthew C. Perry, 2:305–332. Washington, D.C.: B. Tucker, 1856–1857.

———. "Diagnostic Characters of New Species of Phaenogamous Plants Collected in Japan by Charles Wright, with Observations upon the Relations of the Japanese Flora to That of North America and of Other Parts of the Northern Temperate Zone." *Memoirs of the American Academy of Arts and Sciences,* new ser., 6 (1859): 377–452.

———. "Extract from the Concluding Part of a Memoir on the Botany of Japan, in Its Relations to That of North America, and of Other Parts of the Northern Temperate Zone." *American Journal of Science and the Arts,* second ser., 28 (1859): 187–200. Title varies with "On the Botany of Japan, and Its Relation to North America, etc."

Haga Tōru. "Dodonaeus and Tokugawa Culture: Hiraga Gennai and Natural History in Eighteenth-Century Japan." In *Dodonaeus in Japan: Translation and the Scientific Mix in the Tokugawa Period,* edited by W. F. vande Walle and Kasaya Kazuhiko, 241–255. Leuven, Belgium: Leuven University Press, 2001.

Hagino Toshio. *Chōsen Manshū Taiwan ringyō hattatsushiron.* Tokyo: Rin'ya Kōsakai, 1964.

Hanson, Marta E. *Speaking of Epidemics in Chinese Medicine: Disease and the Geographic Imagination in Late Imperial China.* New York: Routledge, 2011.

Hara Hiroshi. *Catalogue of the Works of Dr. Hiroshi Hara.* Tokyo: Yasaka Shobō, 1991.

———. *Distribution Maps of Flowering Plants in Japan* 2. Tokyo: Inoue Book Company, 1959.

———. *List of Plants Common to Japan and North America (Excluding Aleutian).* Tokyo: 1940. Ms. in Harvard University Botany Libraries.

———. *Observationes ad Plantas Asiae Orientalis, I–XX.* Tokyo: Tsumura Kenkyūjo, 1933–1941.

———. "Patterns of Differentiation in Flowering Plants." In *Floristics and Paleofloristics of Asia and Eastern North America,* edited by Alan Graham, 55–60. Amsterdam: Elsevier, 1972.

———. "Preliminary Report on the Flora of Southern Hidaka, Hokkaido." *Botanical Magazine, Tokyo* 48:574–576 (1934): 685–687.

———. *A Revision of Caprifoliaceae of Japan with Reference to Allied Plants in Other Districts and the Adoxaceae.* Tokyo: Academia Scientific Book, 1983.

———. "Takenoshin Nakai: 1882–1952." *Botanical Magazine* 66:775–776 (1953): 1–4.

Harada Hiroshi, ed. *Bōsō no shokubutsushi.* Chiba: Chiba Kenritsu Chūō Hakubutsukan, 2002.

Hardacre, Helen. *Shinto: A History.* New York: Oxford University Press, 2017.

Havens, Thomas R. H. *Nishi Amane and Modern Japanese Thought.* Princeton, N.J.: Princeton University Press, 1970.

———. *Parkscapes: Green Spaces in Modern Japan.* Honolulu: University of Hawai'i Press, 2011.

Hayami Natsuki and Sachi Sri Kantha. "Nobel Prizes for Research in Plant Science: Past, Present and Future." *Reviews in Agricultural Science* 5 (2017): 83–99.

Hayata Bunzō. *The Botany of Mt. Fuji.* Tokyo: Pan-Pacific Science Congress, Japan, 1926.

———. *Flora Montana Formosa: An Enumeration of Plants Found on Mt. Morrison, the Central Chain and Other Mountainous Regions of Formosa at Altitudes of 3,000–13,000 ft.* Tokyo: Tokyo Imperial University, 1908.

———. *Icones Plantarum Formosanarum.* 10 vols. Taihoku: Sōtokufu, 1910–1921.

———. *The Vegetation of Mt. Fuji.* Tokyo: Maruzen, 1911.

Healey, B. J. *The Plant Hunters.* New York: Scribner, 1975.
Henry, Augustine. *A List of Plants from Formosa.* Tokyo: Asiatic Society of Japan, 1896.
Herries, Amanda. "Early Plant-Collectors and Plant-Hunters in Japan." In *A Garden Bequest—Plants from Japan,* edited by Anne Farrar, 1–3. London: Japan Society, 2001.
Hirano, Yukihiro. "Public and Private Support of Basic Research in Japan." *Science,* new series 258:5082 (October 23, 1992): 582–583.
Hirohito, Emperor of Japan. "Linné and Taxonomy in Japan: On the 300th Anniversary of His Birth." In *The Linnaean Legacy: Three Centuries after His Birth,* edited by Mary J. Morris and Leonie Berwick, 115–120. Oxford: Wiley-Blackwell, 2008.
Hirose Hiroyuki and Yamagishi Takaaki, eds. *Nihon tansuisō zukan.* Tokyo: Uchida Rōkakuho Shinsha, 1977.
Hodgson, Christopher P. *A Residence at Nagasaki and Hakodate in 1859–1861, with an Account of Japan Generally.* London: R. Bentley, 1861.
Hollingworth, William. "More Time Needed for Japanese Insects to Tackle Britain's Knotweed Problem." *Japan Times* online, May 3, 2015.
Honda Masaji. *Monographia Poacearum Japonicarum, Bambusoideis Exclusis.* Tokyo: Tokyo Imperial University, 1930.
Hooker, William Jackson. "A Catalogue of Japanese Plants." In *A Residence at Nagasaki and Hakodate in 1859–1861, with an Account of Japan Generally,* by Christopher Pemberton Hodgson, 327–350. London: R. Bentley, 1861.
Horikawa, Yoshiwo. *Atlas of the Japanese Flora: An Introduction to Plant Sociology of East Asia.* 2 vols. Tokyo: Gakken, 1972–1976.
———. *Distributional Studies of Bryophytes in Japan and the Adjacent Regions.* Hiroshima: Hiroshima University, 1955.
Hotta Mitsuru. *Nihon rettō no shokubutsu.* Osaka: Hoikusha, 1974.
Hotta Teikichi. *Taxonomical Studies on the Morus Plants and Their Distribution in Japan and Its Vicinities.* Tokyo: Japan Society for the Promotion of Science, 1958.
Hsu, Jen. "Late Cretaceous and Cenozoic Vegetation in China, Emphasizing Their Connections with North America." *Annals of the Missouri Botanical Garden* 70 (1983): 490–508.
Hu, H. H. "How *Metasequoia,* the 'Living Fossil,' Was Discovered in China." *Journal of the New York Botanical Garden* 49 (1948): 201–207.
Hu Wenqing. *Taiwan de kungyuan.* Taibei: Yuanzu Wenhua, 2007.
Ichimura Tsutsumi. *Nihon yakuyō shokubutsu zufu.* Tokyo: Nihon Yakuhōsha, 1932.
Iinuma Yokusai. *Sōmoku zusetsu.* 4 vols. 3rd ed. Tokyo: Seibidō, 1907–1912.
Imahori Kōzō, ed. *Outline of Plant Science Activities in Japan.* Tokyo: Organizing Committee of the XV International Botanical Congress, 1993.
International Botanical Congress. *International Code of Botanical Nomenclature (Vienna Code).* Koenigstein, Ger.: A. R. G. Gantner Verlag, 2006.
———. *International Code of Nomenclature for Algae, Fungi, and Plants (Melbourne Code).* Koenigstein, Ger.: Koeltz Scientific Books, 2012.
Ishidoya Tsutomu. "Kita Ajia no shokubutsu shigen." In *Tairiku bunka kenkyū,* edited by Keijō Teikoku Daigaku Tairiku Bunka Kenkyūkai, 457–490. Tokyo: Iwanami Shoten, 1940.
Ishiyama Yoshikazu. *Shīboruto: Nihon no shokubutsu ni kaketa shōgai.* Tokyo: Ribun Shuppan, 2000.
Itabashi Akatsuka Botanical Garden Green Advisory Service, Library. Tokyo: Akatsuka Botanical Garden, 2017.
Itō Ihee Sannojō. *A Brocade Pillow: Azaleas of Old Japan.* Translated by Katō Kaname. New York: Weatherhill, 1984.

Itō Kōji. "Hokkaido tōbu enshitchi shokubutsu gunraku no kenkyū." *Hokkaido Daigaku Shokubutsuen kenkyū hōkoku* 1:9 (1963): 1–101.

Itō Tokutarō. "On the History of Botany in Japan." *Journal of Botany* 25 (1887): 1–5.

Itoigawa Junji. *Nihon no shizenshi hakubutsukan.* Tokyo: Tokyo Daigaku Shuppankai, 1993.

Ivings, Steven Edward. *Colonial Settlement and Migratory Labour in Karafuto, 1905–1941.* Ph.D. diss., London School of Economics and Political Science, 2014.

Iwasaki Kan'en. *Bukō sanbutsu shi: Bukō ryakuzu.* Reprint ed. Tokyo: Inoue Shoten, 1967.

———. *Honzō zufu.* 93 vols. Reprint ed. Tokyo: Honzō Zufu Kankōkai, 1916–1922.

Iwashina Tsukasa and Ebihara Atsushi, eds. *Uocchingu Nihon no koyū shokubutsu.* Hadanoshi, Kanagawaken: Tōkai Daigaku Shuppankai, 2014.

Iwatsuki Kunio. *Nihon no shokubutsuen.* Tokyo: Tokyo Daigaku Shuppankai, 2004.

Iwatsuki Kunio, Yamazaki Takashi, David E. Boufford, and Ōba Hideaki, eds. *Flora of Japan.* 6 vols. Tokyo: Kōdansha, 1993–2006.

Iwatsuki Zennosuke. "Two Interesting *Fissidens* Species Common to Japan and Eastern North America." *Journal of Japanese Botany* 33:8 (1958): 245–250.

Japan Association of Botanical Gardens. *Botanical Gardens in Japan.* Tokyo: Japan Association of Botanical Gardens, 1964.

"Japan Boosts Research Funding to Keep Pace with China and South Korea." *Enago* online, May 21, 2018.

"Japan Tests Nuclear-Powered Crops." *Economist* online, May 8, 2012.

"Japanese Researchers Say Science-Budget Hike Isn't Enough." *Nature* online, February 6, 2018.

Jindai Botanical Garden. Tokyo: Tokyo Metropolitan Government, 1962.

Johnson, Jesse Harlan, and Kenji Konishi. *Studies of Mississippian Algae.* Golden: Colorado School of Mines, 1956.

Jonas, Patricia. "Japanese Inspiration." In *Japanese-Inspired Gardens,* edited by Patricia Jonas, 4–7. New York: Brooklyn Botanic Garden, 2001.

———. "Plants for the Japanese-Inspired Garden." In *Japanese-Inspired Gardens,* edited by Patricia Jonas, 66–98. New York: Brooklyn Botanic Garden, 2001.

———, ed. *Japanese-Inspired Gardens.* New York: Brooklyn Botanic Garden, 2001.

Jones, Martin K. "Preface." In *Ancient Plants and People: Contemporary Trends in Archaeobotany,* edited by Marco Madella, Carla Lancelotti, and Manon Savard, ix–xii. Tucson: University of Arizona Press, 2014.

Kachi Naoki. *Population Dynamics and Life History Strategy of Oenothera Erythrosepala in a Sand Dune.* Ph.D. diss., University of Tokyo, 1983.

Kaempfer, Engelbert. *The History of Japan; Together with a Description of the Kingdom of Siam, 1690–92.* 3 vols. Translated by J. G. Scheuchzer. Glasgow: J. MacLehose & Sons, 1906. Reprint of Scheuchzer's 1727 two-volume translation of Kaempfer's manuscript *De beschryving van Japan;* the manuscript was published posthumously in Dutch in 1729 and in German fifty years later.

———. *Kaempfer's Japan—Tokugawa Culture Observed,* edited and translated by Beatrice M. Bodart-Bailey. Honolulu: University of Hawai'i Press, 1999.

Kamimura Minoru. *Makino Tomitarō den.* Osaka: Rokugatsusha, 1955.

Kanai Hiroo. *Distribution Maps of Flowering Plants in Japan.* Tokyo: Inoue Book Company, 1958.

Kanehira Ryōzō. *Nan'yō guntō shokubutsushi.* Tokyo: Inoue Shoten, 1933.

Kankyōchō, ed. *Nihon no zetsumetsu no osore no aru yasei seibutsu: Reddo dēta bukku.* 9 vols. Tokyo: Shizen Kankyō Kenkyū Sentā, 2000–2006.

Kapoor, Maya L. "Conservationists Give Assisted Migration a Second Look." *High Country News* online, August 20, 2018.
Kasaya Kazuhiko. "The Tokugawa Bakufu's Policies for the National Production of Medicine and Dodonaeus' Cruijdeboeck." In *Dodonaeus in Japan: Translation and the Scientific Mix in the Tokugawa Period,* edited by W. F. vande Walle and Kasaya Kazuhiko, 167–186. Leuven, Belgium: Leuven University Press, 2001.
Kawakami Sachio. *Koishikawa shokubutsuen.* Tokyo: Gōgakusha, 1981.
Kihara Hitoshi. *Reimeiki Nihon no seibutsushi.* Tokyo: Yōkendō, 1972.
Kim, Hoi-eun. *Doctors of Empire: Medical and Cultural Encounters between Imperial Germany and Meiji Japan.* Toronto: University of Toronto Press, 2014.
Kimura Kōichi. *Nihon no yakuyō shokubutsu.* Tokyo: Hirokawa Shoten, 1962.
Kimura Yōjirō. "C. P. Thunberg and the Japanese Botany." In *Bicentenary Celebration of C. P. Thunberg's Visit to Japan,* 9–16. Tokyo: Royal Swedish Embassy and Botanical Society of Japan, 1977.
Kimura Yōjirō and Kitamura Shirō, trans. *A Guide to Flora Japonica [of] Philipp Franz von Siebold.* New York: Johnson Reprint Corp., 1977.
Kira Tatsuo. *Nihon no shinrin to bunka: Satoyamaron e no shiten.* Tokyo: Shinjusha, 2011.
Klinkenborg, Verlyn. "Hey, You Calling Me an Invasive Species?" *New York Times,* September 8, 2013, SR10.
Knight, Sam. "The Day of the Knotweed." *Harper's Magazine* online, May 2015.
Kobayashi Yoshio and Shimizu Daisuke. "Cordyceps Species from Japan." *Bulletin of the National Science Museum,* ser. B (Botany) 4:2 (1978): 43–63.
Kobe Municipal Arboretum. *List of Trees and Shrubs.* Kobe: Kobe Municipal Arboretum, 1966.
Kōchi Shinbunsha, ed. MAKINO. Tokyo: Hokuryūkan, 2014.
Kodansha Encyclopedia of Japan. 9 vols. Tokyo: Kōdansha, 1983.
Kokuritsu Kagaku Hakubutsukan, ed. *Kokuritsu Kagaku Hakubutsukan hyakunenshi.* Tokyo: Kokuritsu Kagaku Hakubutsukan, 1977.
———. *Minakata Kumagusu: Hyakunen hayakatta chi no hito.* Tokyo: Kokuritsu Kagaku Hakubutsukan, 2017.
———. *Tsukuba Jikken Shokubutsuen.* Tokyo: Kokuritsu Kagaku Hakubutsukan, 2016.
Kokuritsu Kagaku Hakubutsukan Tsukuba Jikken Shokubutsuen. *Shokubutsuen no arukikata.* Tsukuba: Tsukuba Jikken Shokubutsuen, 2014.
Kolbert, Elizabeth. *The Sixth Extinction: An Unnatural History.* New York: Henry Holt, 2014.
Konishi Kenji. "Anatolipora, a New Dasycladacean Genus, and Its Algal Associates from the Lower Carboniferous of Japan." In *Studies of Mississippian Algae,* edited by Jesse Harlan Johnson and Kenji Konishi, 109–127. Golden: Colorado School of Mines, 1956.
Kowenhoven, Arlette, and Matthi Forrer. *Siebold and Japan: His Life and Works.* Translated by Mark Poysden. Leiden: Hotei, 2000.
Koyama, Tetsuo. *Available Knowledge on the Flora of Japan and the Neighbouring Region.* Toronto: University of Toronto Press, 1961.
———. *Plant Collecting in Japan and Its Adjoining Territories by the Black Ships.* Fujisawa: Nihon University College of Agriculture and Veterinary Medicine, 1994.
Kudō Yūshun. "The Vegetation of Yezo." *Japanese Journal of Botany* 2:4 (1925): 202–292.
Kuginuki Fuji. *SVRS Values: Theories for the Statistical Analysis of Epidermal System Tissue Using the Laminae of Woody Plants.* Tokyo: Kōdansha Scientific, 2000.
Kurata Satoru, ed. *A Bibliography of Forest Botany in Japan, 1940–1963.* Tokyo: University of Tokyo Press, 1966.

Kurata Satoru and Nakaike Toshiyuki, eds. *Nihon no shida shokubutsu zukan*. 9 vols. Tokyo: Tokyo Daigaku Shuppankai, 1979–1997.

Kuriyama Shigehisa. "The Geography of Ginseng and the Strange Alchemy of Needs." In *The Botany of Empire in the Long Eighteenth Century*, edited by Yota Batsaki, Sarah Burke Cahalan, and Anatole Tchikine, 61–72. Washington, D.C.: Dumbarton Oaks Research Library and Collection, 2017.

L. Boehmer and Company. *Catalogue of Japanese Plants, Bulbs, and Seeds*. 4th ed. Yokohama: L. Boehmer, 1903.

Lee, Jung. "Mutual Transformation of Colonial and Imperial Botanizing? The Intimate yet Remote Collaboration in Colonial Korea." *Science in Context* 29:2 (2016): 179–211.

———. "Provincialising Global Botany." In *Worlds of Natural History*, edited by H. A. Curry, N. Jardine, J. A. Secord, and E. C. Spary, 433–446. Cambridge, Eng.: Cambridge University Press, 2018.

Leicht-Young, Stacey A. "Seeing the Lianas in the Trees: Woody Vines of the Temperate Zone." *Arnoldia* 72:1 (2014): 2–12.

Levick, Melba, and Kendall H. Brown. *Japanese-Style Gardens of the Pacific West Coast*. New York: Rizzoli, 1999.

Levy-Yamamori, Ran, and Gerald Taaffe. *Garden Plants of Japan*. Portland, Ore.: Timber Press, 2004.

Lewis, Martin W., and Kären Wigen. *The Myth of Continents: A Critique of Metageography*. Berkeley: University of California Press, 1997.

Lexington, Massachusetts, Conservation Division. *A Trail Guide to Invasive Plants in Lincoln Park*. Lexington, Mass.: Conservation Division, 2016.

Li, H. "Floristic Relationships between Eastern Asia and Eastern North America." *Transactions of the American Philosophical Society*, new series 42 (1952): 371–429.

Li, Yi, Zongming Cheng, William A. Smith, Donna R. Ellis, Yongqin Chen, Xuelian Zheng, Yan Pei, Keming Luo, Degang Zhao, Quanhong Yao, Hui Duan, and Qi Li. "Invasive Ornamental Plants: Problems, Challenges, and Molecular Tools to Neutralize Their Invasiveness." *Critical Reviews in Plant Sciences* 23:5 (2004): 381–389.

A List of Plants and Seeds, Wanted from China and Japan; to Which Is Added, Directions for Bringing Them to Europe. London, 1789.

Lock, Margaret M. *East Asian Medicine in Urban Japan: Varieties of Medical Experience*. Berkeley: University of California Press, 1980.

Madella, Marco, Carla Lancelotti, and Manon Savard. "Introduction." In *Ancient Plants and People: Contemporary Trends in Archaeobotany*, edited by Marco Madella, Carla Lancelotti, and Manon Savard, 3–6. Tucson: University of Arizona Press, 2014.

———, eds. *Ancient Plants and People: Contemporary Trends in Archaeobotany*. Tucson: University of Arizona Press, 2014.

Makino Tomitarō. *Makino shokubutsu zukan*. Tokyo: Hokuryūkan, 1940.

———. *Makino Tomitarō shokubutsuki*. Tokyo: Sakurai Shoten, 1948.

———. "Preface." In Iinuma Yokusai, *Sōmoku zusetsu*, 1:1–15. 3rd ed. Tokyo: Seibidō, 1907–1912.

Maldanado Torres, Nelson. *Against War: Views from the Underside of Modernity*. Durham, N.C.: Duke University Press, 2008.

Manchester, Steven R. "Biogeographical Relationships of North American Tertiary Floras." *Annals of the Missouri Botanical Garden* 86:2 (1999): 472–522.

Manchester, Steven R., Zhi-duan Chen, An-ming Lu, and Kazuhiko Uemura. "Eastern Asian

Endemic Seed Plant Genera and Their Paleogeographic History throughout the Northern Hemisphere." *Journal of Systematics and Evolution* 47:1 (2009): 1–42.

Manke, Kara. "Is Habitat Restoration Actually Killing Plants in the California Wildlands?" *Berkeley News* online, January 2, 2019.

Marcon, Federico. "Inventorying Nature: Tokugawa Yoshimune and the Sponsorship of Honzōgaku in Eighteenth-Century Japan." In *Japan at Nature's Edge: The Environmental Context of a Global Power,* edited by Ian Jared Miller, Julia Adeney Thomas, and Brett L. Walker, 189–206. Honolulu: University of Hawai'i Press, 2013.

———. *The Knowledge of Nature and the Nature of Knowledge in Early Modern Japan.* Chicago: University of Chicago Press, 2015.

Masamune Genkei. *Floristic and Geobotanical Studies on the Island of Yakushima, Osumi Province.* Taihoku: Taihoku Imperial University, 1934.

———. *Kainantō shokubutsushi.* Taihoku : Taiwan Sōtokufu, 1944.

Masuda Yoshio. *Wasurerareta shokubutsu gakusha: Nagamatsu Atsusuke no karei na tenshin.* Tokyo: Chūō Kōronsha, 1987.

Matsui Ryūgo and Iwasaki Masashi, eds. *Minakata Kumagusu no mori.* Kyoto: Hōjōdō, 2005.

Matsumura Jinzō, ed. *Nihon shokubutsi meii.* Tokyo: Maruzen, 1884.

Matsunaga Waki. *Shokubutsu de mirai o tsukuru.* Kyoto: Kagaku Dōjin, 2008.

Matsuura Hajime. *A Bibliographical Monograph on Plant Genetics (Genic Analysis), 1900–1925.* Tokyo: Tokyo Imperial University, 1929.

McDonald, Kate. *Placing Empire: Travel and the Social Imagination in Imperial Japan.* Oakland: University of California Press, 2017.

Merrill, Elmer Drew. "Cooperative Botanical Exploration in Eastern Asia." *Science Counselor* 11:1 (1948): 15–16, 35–37.

Merrill, Elmer Drew, and Egbert Hamilton Walker. *A Bibliography of Eastern Asiatic Botany.* Rev. ed. Jamaica Plain, Mass.: Arnold Arboretum of Harvard University, 1938.

Métailié, Georges. "Comparative Study of the Introduction of Modern Botany in Japan and China." *Historia Scientiarum* 11:3 (2002): 205–217.

Meyer, Robinson. "Geology's Timekeepers Are Feuding." *Atlantic* online, July 20, 2018.

Miki Shigeru. "Plant Fossils from the Stegodon Beds and the Elephas Beds Near Akashi." *Japanese Journal of Botany* 8:4 (1937): 303–341.

Miller, Ian Jared. *The Nature of the Beasts: Empire and Exhibition at the Tokyo Imperial Zoo.* Berkeley: University of California Press, 2013.

Miller, Ian Jared, Julia Adeney Thomas, and Brett L. Walker, eds. *Japan at Nature's Edge: The Environmental Context of a Global Power.* Honolulu: University of Hawai'i Press, 2013.

Miller-Rushing, Abraham J. "Impacts of Climate Change on the Phenology of Temperate Forest Plants and Birds in Massachusetts and Japan." Ph.D. diss., Boston University, 2007.

Milne, Richard Ian. "Northern Hemisphere Plant Disjunctions: A Window on Tertiary Land Bridges and Climate Change?" *Annals of Botany* 98:3 (2006): 465–472.

Mitford, Algernon Freeman. *The Bamboo Garden.* London: Macmillan, 1896.

———. *Mitford's Japan: Memories and Recollections, 1866–1906.* Edited by Hugh Cortazzi. London: Athlone, 1995.

———. *Tales of Old Japan.* 2 vols. London: Macmillan, 1871.

Miyabe Kingo and Kudō Yūshun, eds. *Hokkaido shuyō jumoku zufu.* 28 vols. Tokyo: Sanshūsha, 1920–1931.

Miyachi Shigetō. "Preface." *Nihon shizen kagaku shūhō* 7 (1982): 1.

Miyake Kiichi. "The Botanical Laboratory and the Botanical Garden of the Tokyo Imperial University." *Journal of Applied Microscopy* 4:10 (1901): 1477–1481.

Miyashita-Theado, Kaori. "A Study of Kunikida Doppo: 'Musashino' and Three Short Stories." Master's thesis, University of Colorado, 2010.
Miyawaki Akira, ed. *Nihon shokuseishi.* 10 vols. Tokyo: Shibundō, 1980–1989.
Miyawaki Akira, Iwatsuki Kunio, and Miroslav M. Grandtner, eds. *Vegetation in Eastern North America: Vegetation System and Dynamics under Human Activity in the Eastern North American Cultural Region in Comparison with Japan.* Tokyo: University of Tokyo Press, 1994.
Miyoshi Manabu. "Hiraseshi 'Ichō seichū' ni oite." *Shokubutsugaku zasshi: The Botanical Magazine, Tokyo* 10:12 (1896): 409–415.
Miyoshi Manabu and Makino Tomitarō. *Pocket-Atlas of Alpine Plants of Japan.* 2 vols. 2nd rev. ed. Tokyo: Seibidō, 1907.
Mizuno Hiromi. *Science for the Empire: Scientific Nationalism in Modern Japan.* Stanford: Stanford University Press, 2009.
Morita Mitsuyoshi, ed. *Kika shokubutsu no shizenshi—shinryaku to kōran no seitaigaku.* Sapporo: Hokkaido Daigaku Shuppankai, 2012.
Morris, Mary J., and Leonie Berwick, eds. *The Linnaean Legacy: Three Centuries after His Birth.* Oxford: Wiley-Blackwell, 2008.
Mueggler, Erik. *The Paper Road: Archive and Experience in the Botanical Exploration of West China and Tibet.* Berkeley: University of California Press, 2011.
Muntschick, Wolfgang. "The Plants That Carry His Name: Kaempfer's Study of the Japanese Flora." In *The Furthest Goal: Engelbert Kaempfer's Encounter with Tokugawa Japan,* edited by Beatrice Bodart-Bailey and Derek Massarella, 71–95. Richmond, Eng.: Curzon Press, 1995.
Murasaki Nagaaki. *Taihoku shashinchō.* Taihoku: Shinkōdō Shoten, 1913.
Nagai Hiroyuki, Nukaga Yoshio, Saeki Kōji, and Akabayashi Akira. "Self-Regulation of Recombinant DNA Technology in Japan in the 1970s." *Historia Scientiarum* 19:1 (2009): 1–18.
Nagakubo Hen'in. *Matsumura Jinzō no shōgai.* Tokyo: Akatsuki Inshokan, 1997.
Nagao Hideyuki and Fukiharu Toshimitsu. "Fungal Flora in Chiba Prefecture." In *Bōsō no shokubutsushi,* edited by Harada Hiroshi, 131–132. Chiba: Chiba Kenritsu Chūō Hakubutsukan, 2002.
Nakai Takenoshin. *Flora Koreana.* 2 vols. Tokyo: Tokyo Imperial University, 1909–1911.
Nakamura Futoshi. "Current Status and Nationwide Database of the River and Floodplain Ecosystem in Japan." In *The Biodiversity Observation Network in the Asia-Pacific Region,* edited by Nakano Shin'ichi, Yahara Tetsukazu, and Nakashizuka Tōru, 229–245. London: Springer, 2012.
Nakano Harufusa. *Shokubutsu seiri oyobi seitaigaku jikkenhō.* Tokyo: Shōkabō, 1933.
Nakano Shin'ichi, Yahara Tetsukazu, and Nakashizuka Tōru, eds. *The Biodiversity Observation Network in the Asia-Pacific Region.* London: Springer, 2012.
Nakazawa Shin'ichi. *Mori no barokku.* Tokyo: Serika Shobō, 1992.
Nance, R. Damian, and J. Brendan Murphy. *Supercontinents and the Case for Pannotia.* London: Geological Society, Special Publications, March 1, 2018. P.SP470.5.
Nappi, Carla Suzan. *The Monkey and the Inkpot: Natural History and Its Transformation in Early Modern China.* Cambridge, Mass.: Harvard University Press, 2009.
Neilson, Ronald P., Louis F. Pitelka, Allen M. Solomon, Ran Nathan, Guy F. Midgley, Jose M. V. Fragoso, Heike Lischke, and Ken Thompson. "Forecasting Regional to Global Plant Migration Response to Climate Change." *Bioscience* 55:8 (2005): 749–759.
Nelson, E. Charles. "So Many Really Fine Plants—An Epitome of Japanese Plants in Western European Gardens." *Curtis's Botanical Magazine* 16:2 (1999): 52–68.

Nihon Shokubutsu Bunrui Gakkai Kenritsu Kinen Kokusai Shinpojiumu. *Nihonsan shokubutsu o meguru seibutsu chiri.* Kyoto: Nihon Shokubutsu Bunrui Gakkai, 2001.

Nihon Shokubutsuen Kyōkai. *Nihon no shokubutsuen.* Tokyo: Nihon Shokubutsuen Kyōkai, 1987.

Nihon Shokubutsu Gakkai, ed. *Shokubutsugaku no hyakka jiten.* Tokyo: Maruzen, 2016.

Nihon Shokubutsu Gakkai Hyakunenshi Henshū Iinkai, ed. *Nihon no shokubutsugaku hyakunen no ayumi.* Tokyo: Nihon Shokubutsu Gakkai, 1982.

Nippon Shinyaku Institute for Botanical Research. *List of Plants.* Kyoto: Nippon Shinyaku Institute for Botanical Research, 1963.

Nishikawa Tsunehiko, ed. *Chromosome Atlas of Flowering Plants in Japan.* Tokyo: National Museum of Nature and Science, 2008.

Nitzelius, Tor. *The Nordic Arboretum Expedition to Japan 1976.* Alnarp, Sweden: Department of Dendrology, 1978.

Novacek, Michael J. "Prehistory's Brilliant Future." *New York Times,* November 9, 2014, SR6–SR7.

Numata Makoto. *The Flora and Vegetation of Japan.* Tokyo: Kodansha, 1974.

Nuttall, Thomas. *The Genera of North American Plants, and a Catalogue of the Species, to the Year 1817.* Philadelphia: D. Heartt, 1818.

Ōba Hideaki, ed. *Nihon no kika shokubutsu zufu.* Kamakura: Abokkusha, 2009.

———, ed. *Nihon no zetsumetsu kigu shokubutsu zufu.* Kamakura: Apokkusha, 2004.

———, ed. *Nihon shokubutsu kenkyū no rekishi: Koishikawa Shokubutsuen 300nen no ayumi.* Tokyo: Tokyo Daigaku Shuppankai, 1996.

———, ed. *Shokubutsu bunka jinbutsu jiten.* Tokyo: Nichigai Asoshiētsu, 2007.

Ōba Hideaki and David E. Boufford, eds. *The Botanical Collections: Proceedings of the Symposium "Siebold in the 21st Century" Held at the University Museum, the University of Tokyo, in 2003.* Tokyo: University of Tokyo University Museum, 2005.

Ōfuna Botanical Garden. *Kanagawa Prefectural Ofuna Botanical Garden.* Kamakura: Kanagawaken, 1970.

Ogawa-Onishi, Yuko, Pam M. Berry, and Nobuyuki Tanaka. "Assessing the Potential Impacts of Climate Change and Their Conservation Implications in Japan: A Case Study of Conifers." *Biological Conservation* 143:7 (2010): 1728–1736.

Ogilvie, Brian W. *The Science of Describing: Natural History in Renaissance Europe.* Chicago: University of Chicago Press, 2006.

Ōhashi Hiroyoshi. "Bunzō Hayata and His Contributions to the Flora of Taiwan." *Taiwania* 54:1 (2009): 1–27.

———. *List of the Type Specimens in the Herbaria of Japan.* Kyoto: Working Group on Vascular Plants, 1981.

Ohnuki-Tierney, Emiko. *Flowers That Kill: Communicative Opacity in Political Spaces.* Stanford: Stanford University Press, 2015.

Ōi [Ohwi] Jisaburō. *The Flora of Japan.* Edited by Frederick G. Meyer and Egbert H. Walker. Washington, D.C.: Smithsonian Institution, 1965.

Okita Yoshihiro and J. Leland Hollenberg. *The Miniature Palms of Japan.* New York: Weatherhill, 1981.

Oliver, Daniel. "Notes upon a Few of the Plants Collected, Chiefly Near Nagasaki, Japan, and in the Islands of the Korean Archipelago, in the Years 1862–63, by Mr. Richard Oldham, Late Botanical Collector Attached to the Royal Gardens, Kew." *Journal of the Linnean Society* 8 (1867): 163–170.

Ōnuki Shigeru. *Man'yō shokubutsu jiten.* Tokyo: Kureo, 1998.

Ōtsuka Keisetsu and Yakazu Dōmei, eds. *Matsuoka Joan*. Tokyo: Meicho Shuppan, 1980.
Parmesan, Camille, and Gary Yohe. "A Globally Coherent Fingerprint of Climate Change Impacts across Natural Systems." *Nature* 421 (2003): 37–42.
Parsons, Christopher M. "The Natural History of Colonial Science: Joseph-François Lafitau's Discovery of Ginseng and Its Afterlives." *William and Mary Quarterly*, 3rd ser., 73:1 (2016): 37–72.
———. "Wildness without Wilderness: Biogeography and Empire in Seventeenth-Century French North America." *Environmental History* 22:4 (2017): 643–667.
Pearson, Lisa. *Arnold Arboretum*. Charleston, S.C.: Arcadia Publishing, 2016.
———. "John George Jack: Dendrologist, Educator, Plant Explorer." *Arnoldia* 71:4 (2014): 2–11.
Perry, Matthew C. *Narrative of the Expedition of an American Squadron to the China Seas and Japan: Performed in the Years 1852, 1853, and 1854, under the Command of Commodore M. C. Perry, United States Navy, by Order of the Government of the United States*. 3 vols. Washington, D.C.: B. Tucker, 1856–1857.
Phillips, Nicky. "Japanese Research Leaders Warn about National Science Decline." *Nature* online, October 20, 2017.
Plutschow, Herbert E. *Philipp Franz von Siebold and the Opening of Japan: A Re-evaluation*. Folkestone, Eng.: Global Oriental, 2007.
Pollan, Michael. "The Intelligent Plant: Scientists Debate a New Way of Understanding Flora." *New Yorker*, December 23–30, 2013, pp. 92–96, 98–102, 104–105.
Quammen, David. "The Scientist Who Scrambled Darwin's Tree of Life." *New York Times* online, August 13, 2018.
Reitan, Richard. "Ecology and Japanese History: Reactionary Environmentalism's Troubled Relationship with the Past." *Asia-Pacific Journal* online, February 1, 2017.
Rietbergen, Peter. "Meeting Point Deshima: Scholarly Communication between Japan and Europe up till around 1800." In *Empire and Science in the Making: Dutch Colonial Scholarship in Comparative Global Perspective, 1760–1830*, edited by Peter Boomgaard, 207–230. New York: Palgrave Macmillan, 2013.
Rix, Martyn. *Golden Age of Botanical Art*. London: Andre Deutsch, 2012.
Ryall, Julian. "Japanese Scientists Create Cherry Tree That Blossoms All Year Round." *Telegraph* online, February 17, 2010.
Sagers, John H. *Origins of Japanese Wealth and Power: Reconciling Confucianism and Capitalism, 1830–1885*. New York: Palgrave Macmillan, 2006.
Saraiva, Tiago. *Fascist Pigs: Technoscientific Organisms and the History of Fascism*. Cambridge, Mass.: MIT Press, 2016.
Sargent, Charles Sprague. *Forest Flora of Japan: Notes on the Forest Flora of Japan*. Boston, Mass.: Houghton, Mifflin, 1894.
———. "The Forest: Some Japanese Trees in New England." *Garden and Forest* 2:89 (1889): 537–538.
Satō Yōichirō. "Rice of Asian Origin." In *Ancient Plants and People: Contemporary Trends in Archaeobotany*, edited by Marco Madella, Carla Lancelotti, and Manon Savard, 265–276. Tucson: University of Arizona Press, 2014.
Savatier, Paul Amedée Ludovic. "On the Increase of the Flora of Japan." *Transactions of the Asiatic Society of Japan* 2 (1874): 232–243.
Scharf, Sarah Tovah. "Identification Keys, the 'Natural Method,' and the Development of Plant Identification Manuals." *Journal of the History of Biology* 42 (2009): 73–117.
Scharff, Robert Francis. "On the Evidence of a Former Land-Bridge between Northern

Europe and North America." *Proceedings of the Royal Irish Academy. Section B: Biological, Geological, and Chemical Science* 28 (1909/1910): 1–28.
Schiebinger, Londa. *Plants and Empire: Colonial Bioprospecting in the Atlantic World.* Cambridge, Mass.: Harvard University Press, 2004.
Schiebinger, Londa, and Claudia Swan, eds. *Colonial Botany: Science, Commerce, and Politics in the Early Modern World.* Philadelphia: University of Pennsylvania Press, 2005.
Schierenbeck, Kristina A. "Japanese Honeysuckle (*Lonicera japonica*) as an Invasive Species: History, Ecology, and Context." *Critical Reviews in Plant Sciences* 23:5 (2004): 391–400.
Scott, James C. *Seeing Like a State: How Certain Schemes to Improve the Human Condition Have Failed.* New Haven, Conn.: Yale University Press, 1998.
Seaton, Philip, Takayoshi Yamamura, Akiko Sugawa-Shimada, and Kyungjae Jang. *Contents Tourism in Japan: Pilgrimages to "Sacred Sites" of Popular Culture.* Amherst, N.Y.: Cambria Press, 2017.
Seibutsu Gokenkyūjo, ed. *Izu Suzaki no shokubutsu.* Osaka: Hoikusha, 1980.
Seki Tarō, Nakanishi Hiroki, Suzuki Hyōji, and Horikawa Yoshio. *A Flora of Vascular Plants of Itsukushima (Miyajima) Island, Southern Japan.* Hiroshima: Committee for the Urgent Investigation of the Primeval Forest and Scenic Reserves of Itsukushima (Miyajima) Island, 1975.
Sharp, Aaron J. "The Possible Significance of Some Exotic Distribution of Plants Occurring in Japan and/or North America." In *Floristics and Paleofloristics of Asia and Eastern North America,* edited by Alan Graham, 61–64. Amsterdam: Elsevier, 1972.
Shibata Keita. *Tanso oyobi chisso no dōka sayō.* Tokyo: Iwanami, 1931.
Shimizu Tatemi, ed. *Nihon no kika shokubutsu.* Tokyo: Heibonsha, 2003.
Shirahata Yōzaburō. "The Development of Japanese Botanical Interest and Dodonaeus' Role: From Pharmacopeia to Botany and Horticulture." In *Dodonaeus in Japan: Translation and the Scientific Mix in the Tokugawa Period,* edited by W. F. vande Walle and Kasaya Kazuhiko, 263–273. Leuven, Belgium: Leuven University Press, 2001.
Shirai Mitsutarō. "A Brief History of Botany in Old Japan." In Gakujutsu Kenkyū Kaigi, *Scientific Japan, Past and Present,* 213–227. Kyoto: Gakujutsu Kenkyū Kaigi, 1926.
———. *Shokubutsu yōikō.* Tokyo: Oka Shoin, 1926.
Shiraiwa Takumi. *Makino Tomitarō to Kobe.* Kobe: Kobe Shinbun Sōgō Shuppan Sentā, 2008.
Shizen Kankyō Kenkyū Sentā, ed. *Nihon no gairai seibutsu.* Tokyo: Heibonsha, 2008.
Siebold, Philipp Franz von. *Flora Japonica: Sive, Plantae, quas in Imperio Japonico Collegit, Descripsit.* 2 vols. Lugduni Batavorum: The Author, 1835, 1870.
———. *Siebold's Florilegium of Japanese Plants.* Edited by Kimuta Yōjirō and Valerii I. Grubov. Tokyo: Maruzen, 1993.
Smith, Daniel. "Consume, Screw, Kill." *Harper's Magazine,* May 2014, 84–86, 88–89.
Spongberg, Stephen A. *A Reunion of Trees: The Discovery of Exotic Plants and Their Introduction into North American and European Landscapes.* Cambridge, Mass.: Harvard University Press, 1990.
Stearn, William T. "Engelbert Kaempfer (1651–1716)—Pioneer Investigator of Japanese Plants." *Curtis's Botanical Magazine* 16:2 (1999): 103–115.
Sugawara Shigezō. *Karafuto shokubutsu zushi.* 4 vols. Toyohara: Karafuto Shokubutsu Zushi Kankōkai, 1937–1940.
———. *Plants of Saghalien = Karafuto no shokubutsu.* Toyohara, Karafuto: Sakhalin Society for Publication of Botanical Works, 1937.
Suka Takeshi, Okamoto Tōru, and Ushimaru Atsushi. *Kusachi to Nihonjin.* Tokyo: Tsukiji Shokan, 2012.

Sutton, Sylvia Barry. *Charles Sprague Sargent and the Arnold Arboretum*. Cambridge, Mass.: Harvard University Press, 1970.
Suzuki Sadao. *Nihon takeka shokubutsu sōmokuroku*. Tokyo: Gakken, 1982.
Suzuki Shigetaka. *Nihon tokusan shokubutsu zusetsu*. Tokyo: Suzuki Botanical Institute, 1965.
Tachibana Morikuni [Nakamura Yuzeishi]. *Kusubana ōshukubai*. 1727f. Reissued as *Japanese Botany: Being a Facsimile of a Japanese Book*. Philadelphia: J. B. Lippincott, 1850–1855.
Tachibana Setsu. "The 'Capture' of Exotic Natures: Cross-cultural Knowledge and Japanese Gardening in Early 20th Century Britain." *Jinbun chiri—Japanese Journal of Human Geography* 66:6 (2014): 4–18.
Takahashi Hideki and Kawahara Noriko, eds. *Kuraku hakase to Sapporo no shokubutsu*. Sapporo: Hokkaido Daigaku Sōgō Hakubutsukan, 2012.
Takamatsu Masahiko. *The Marine Algae from Matsushima Bay, Miyagi Prefecture, Northeastern Honshū, Japan*. Sendai: Saitō Hoonkai, 1936.
Takatori Jisuke. *Saishiki shaseizu Nihon no yakuyō shokubutsu*. 2 vols. Tokyo: Hirokawa Shoten, 1966–1970.
Takeda Hisayoshi. *Alpine Flowers of Japan*. Tokyo: Sanseidō, 1938.
———. *Systematic List of Economic Plants in Japan*. Tokyo: SCAP, Natural Resources Section, 1949.
Takeda Yakuhin Kōgyō Kabushiku Kaisha. *List of Plants September 1961: The Kyoto Takeda Herbal Garden Attached to the Research Laboratory of Takeda Chemical Industries, Ltd*. Kyoto: Takeda Yakuhin Kōgyō Kabushiki Kaisha, 1961.
Takemi Tarō, ed. *Herbal Medicine: Kampō, Past and Present*. Tokyo: Tsumura Juntendō, 1985.
Tama Shinrin Kagakuen. *Tama Forest Science Garden Study Guide*. Hachiōji: Tama Shinrin Kagakuen, 2016.
Tamura Michio. *Ikiteiru kodai shokubutsu*. Expanded ed. Osaka: Hoikusha, 1974.
Tanai Toshimasa. "Tertiary History of Vegetation in Japan." In *Floristics and Paleofloristics of Asia and Eastern North America*, edited by Alan Graham, 235–255. Amsterdam: Elsevier, 1972.
Tanaka Atsuo. *Shinrin no shinjōshiki*. Tokyo: Yōsensha, 2011.
Tanaka Yoshio. *Yūyō shokubutsu zusetsu*. 7 vols. Tokyo: Dai Nippon Nōkai, 1891.
Tatewaki Misao. *Vegetation of Hokkaido*. Tokyo: Dai Nippon, 1936.
Tatsumi Tatsuo, ed. *The Japanese Scientific Expeditions to Svalbard, 1983–1988*. Tokyo: Kyōikusha, 1990.
Tawara Hiromi. *Makino shokubutsu zukan no nazo*. Tokyo: Heibonsha, 1999.
Thai, Philip. *China's War on Smuggling: Law, Economic Life, and the Making of the Modern State, 1842–1965*. New York: Columbia University Press, 2018.
"Theory Suggests Root Efficiency, Independence Drove Global Spread of Flora." *Princeton Weekly Bulletin* online, February 28, 2018.
Thompson, John N. *The Geographic Mosaic of Coevolution*. Chicago: University of Chicago Press, 2005.
Thornber, Karen Laura. *Ecoambiguity: Environmental Crises and East Asian Literatures*. Ann Arbor: University of Michigan Press, 2012.
———. *Empire of Texts in Motion: Chinese, Korean, and Taiwanese Transculturations of Japanese Literature*. Cambridge, Mass.: Harvard University Asia Center, 2009.
Thornton, Michael. "A Capitol Orchard: Botanical Networks and the Production of Urban Space in Early Meiji Sapporo." Unpublished manuscript, Harvard University, 2014.
Thunberg, Carl Peter. *Flora iaponica: Sistens plantas insularum iaponicarum secundum sistema sexuale emendatum redactas ad XX classes, ordines, genera et species cum differentiis specificis,*

synonymis paucis, descriptionibus concinnis et XXXIX iconibus adiectis. Lipsiae: In Bibliopolio I. G. Mülleriano, 1784.

———. *Japan Extolled and Decried: Carl Peter Thunberg's Travels in Japan, 1775–1776.* Ed. Timon Screech. London: Routledge, 2005.

Tiffney, Bruce H. "Perspectives on the Origin of the Floristic Similarity between Eastern Asia and Eastern North America." *Journal of the Arnold Arboretum* 66 (1985): 73–94.

Tohoku Shokubutsu Seitai Danwakai, ed. *Shokubutsu seitai ronshū.* Sendai: Tohoku Shokubutsu Seitai Danwakai, 1978.

Tokita Jun. *The Marine Algae of Southern Saghalien.* Sapporo: Hokkaido University, 1945.

Tokyo Daigaku Sōgō Kenkyū Shiryōkan. *Catalog of the Fossil Plants Preserved in the Herbarium, Department of Botany, University Museum, University of Tokyo.* Tokyo: University of Tokyo, 1984.

———. *Catalog of the Type Specimens Preserved in the Herbarium of the Department of Botany in the University Museum, University of Tokyo.* 11 vols. Tokyo: University of Tokyo, 1981–2002.

Tokyo Metropolitan Jindai Botanical Park. *List of Plants, 1991.* Tokyo: Jindai Botanical Park, 1991.

Totman, Conrad. *The Green Archipelago: Forestry in Preindustrial Japan.* Berkeley: University of California Press, 1989.

———. *Japan's Imperial Forest Goryōrin, 1889–1946.* Folkestone, Eng.: Global Oriental, 2007.

Totoya Masatsugu, Dirk Spencer, Satoe Sawai-Toyota, Wang Jiaqi, Tong Zhang, Abraham J. Koo, Gregg A. Howe, and Simon Gilroy. "Glutamate Triggers Long-Distance, Calcium-Based Plant Defense Signaling." *Science* 361:6047 (2018): 1112–1115.

Toyokuni Hideo. "On Some Noteworthy Plants from Hokkaido, Japan." *Shinshū Daigaku Kyōyōbu kiyō* 13 (1979): 127–133.

Trambaiolo, Daniel. "Native and Foreign in Tokugawa Medicine." *Journal of Japanese Studies* 39:2 (2013): 299–324.

Tsukamoto Yōtarō and John L. Creech. *Japanese Horticulture: Origins and History.* Yokohama: Woods Press, 2015.

Tsurumi Kazuko. *The Adventure of Ideas: A Collection of Essays on Patterns of Creativity and a Theory of Endogenous Development.* Tokyo: Japanime, 2016. Accessed July 29, 2018. http://www.japanime.com.

———. "Comparative Forms of Creativity among Japanese Folklorists: Origuchi Shinobu, Yanagita Kunio, and Minakata Kumagusu." In Tsurumi Kazuko, *The Adventure of Ideas: A Collection of Essays on Patterns of Creativity and a Theory of Endogenous Development,* 62–101. Tokyo: Japanime, 2016. Accessed July 29, 2018. http://www.japanime.com.

———. "The Minakata-Mandala: A Paradigm Change for the Future." In Tsurumi Kazuko, *The Adventure of Ideas: A Collection of Essays on Patterns of Creativity and a Theory of Endogenous Development,* 165–187. Tokyo: Japanime, 2016. Accessed July 29, 2018. http://www.japanime.com.

Uchida, Jun. *Brokers of Empire: Japanese Settler Colonialism in Korea, 1876–1945.* Cambridge, Mass.: Harvard University Asia Center, 2014.

Uemura Kazuhiko. *Late Miocene Floras in Northeast Honshu, Japan.* Tokyo: National Science Museum, 1988.

Ueno Masuzō. *Hakubutsugaku no jidai.* Tokyo: Yasaka Shobō, 1990.

———. *Nihon hakubutsugakushi.* Tokyo: Heibonsha, 1973.

United States Naval Expedition to Japan (1852–1854). *Drawings for the U.S. Japan Expedition*

and U.S. North Pacific Exploring Expedition, 1852–1856 (inclusive). Unpublished drawings by Isaac Sprague (1811–1895) in the Harvard University Herbaria.
"Unmarked Garden Quarters of China's Last Emperor Puyi May Get Its Place in History." *South China Morning Post* online, June 9, 2014.
Utech, Frederick H. "Vascular Floral Anatomy of Helonias Bullata (Liliaceae-Helonicae), with a Comparison to the Asian Heloniopsis Orientalis." *Annals of the Carnegie Museum* 47:8 (1978): 169–191.
vande Walle, W. F., and Kasaya Kazuhiko, eds. *Dodonaeus in Japan: Translation and the Scientific Mix in the Tokugawa Period.* Leuven, Belgium: Leuven University Press, 2001.
Veitch, James H. *A Traveller's Notes, or Notes of a Tour through India, Malaysia, Japan, Corea, the Australian Colonies and New Zealand during the Years 1891–1893.* Chelsea, Eng.: J. Veitch & Sons, 1896.
Veitch, John Gould. "Notes on the Agriculture, Trees, and Flora of Japan." In *The Capital of the Tycoon: A Narrative of a Three Years' Residence in Japan,* by Rutherford B. Alcock, 2:475–490. London: Longman, Green, Longman, Roberts & Green, 1863.
Voosen, Paul. "NASA Rover Hits Organic Pay Dirt on Mars." *Science* online, June 7, 2018.
Wakabayashi, Bob Tadashi. *Anti-foreignism and Western Learning in Early-Modern Japan: The New Theses of 1825.* Cambridge, Mass.: Council on East Asian Studies, Harvard University, 1986.
Walker, Brett L. *A Concise History of Japan.* Cambridge, Eng.: Cambridge University Press, 2015.
Walravens, Harmut, ed. *A Japanese Herbal in the Wellcome Institute for the History of Medicine: A Contribution to the History of the Transfer of Scientific Knowledge from Europe to Japan.* Wiesbaden, Ger.: Harrassowitz, 2005.
"Want a Nobel Prize? . . . Don't Study Plants." *How Plants Work* online, October 11, 2012.
Warner, Sam Bass, Jr. "Foreword." In *A Reunion of Trees: The Discovery of Exotic Plants and Their Introduction into North American and European Landscapes,* by Stephen A. Spongberg, vii–xi. Cambridge, Mass.: Harvard University Press, 1990.
Watanabe Kōichi. *Moving Plants.* Kyoto: Seigensha, 2015.
Watari-um, ed. *Kumagusu no mori: Minakata Kumagusu no mita uchū.* Tokyo: Shinchōsha, 2007.
Weiner, Michael, ed. *Race, Ethnicity and Migration in Modern Japan.* London: Routledge-Curzon, 2004.
Wen, Jun. "Evolution of Eastern Asian and Eastern North American Disjunct Distributions in Flowering Plants." *Annual Review of Ecology and Systematics* 30 (1999): 421–455.
Whittle, Tyler. *The Plant Hunters.* London: Heinemann, 1970.
Whitwam, Ryan. "Japanese Scientists Use Particle Accelerator to Create Salt-Resistant Rice." Extreme Tech. Accessed May 10, 2014. http://www.extremetech.com.
Williams, Roger L. *Botanophilia in Eighteenth-Century France: The Spirit of the Enlightenment.* Dordrecht, Neth.: Kluwer, 2001.
Wilson, Ernest Henry. *The Cherries of Japan.* Cambridge, Mass.: Harvard University Press, 1916.
———. *The Conifers and Taxads of Japan.* Cambridge, Mass.: Harvard University Press, 1916.
———. *The Lilies of Eastern Asia: A Monograph.* London: Dulau, 1925.
Woese, Carl R., and George E. Fox. "Phylogenetic Structure of the Prokaryotic Domain: The Primary Kingdoms." *Proceedings of the National Academy of Sciences* 74:11 (1977): 5088–5090.
Wright, Charles. *Plant Lists from the North Pacific Exploring Expedition, 1853–1856.* Ms., Harvard University Botany Libraries.

Wu Zhengyi. "On the Significance of Pacific Intercontinental Discontinuity." *Annals of the Missouri Botanical Garden* 70:4 (1983): 577–590.
Wulf, Andrea. *Invention of Nature: Alexander von Humboldt's New World*. New York: Knopf, 2015.
Yahara Tetsukazu, Fujii Shinji, Itō Momomi, and Nagata Yoshio, eds. *Reddo dēta purantsu: Zetsumetsu kigu shokubutsu zukan*. Tokyo: Yama to Keikokusha, 2015.
Yamanaka and Company. *A Collection of Rare and Beautiful Japanese and Arboreal Plants*. Exh. cat. Boston, Mass.: Leonard and Company, Auctioneers, 1899.
Yano Norimichi, ed. *Nihon no shokusei: Shinryaku to kōran no seitaigaku*. Tokyo: Tōkai Daigaku Shuppankai, 1988.
Yatabe Ryōkichi. "A Few Words of Explanation to European Botanists." *Botanical Magazine, Tokyo* 4:44 (1890): 355–359.
———. *Nihon shokubutsu zukai*. 3 parts. Tokyo: Maruzen, 1891–1893.
———. *Nihon shokubutsuhen*. Tokyo: Dai Nihon Tosho, 1900.
Yokohama Ueki Kabushiki Kaisha. *Descriptive Catalogue*. Yokohama: Yokohama Nursery Company, 1901–1924, 1936.
Young, John. *The Research Activities of the South Manchurian Railway Company, 1907–1945: A History and Bibliography*. New York: Columbia University East Asian Institute, 1966.
Young, Louise. *Japan's Total Empire: Manchuria and the Culture of Wartime Imperialism*. Berkeley: University of California Press, 1998.
Yumenoshima Nettai Shokubutsukan. *Yumenoshima Tropical Greenhouse Dome*. Tokyo: Yumenoshima Nettai Shokubutsukan, 2016.
Zimmer, Carl. "For Trees under Threat, Flight May Be Best Response." *New York Times* online, September 18, 2004.

Index

Page numbers in **boldface** refer to illustrations.

About the Author

Thomas R. H. Havens is professor of history at Northeastern University. Trained in history at Princeton and the University of California, Berkeley, in 1960 he first visited Tokyo's Shinjuku district, home of the Shinjuku Imperial Garden in the capital's busiest commercial, government, and transportation complex. Shinjuku and its public areas, as well as the nearby Koishikawa Botanical Gardens, are key sites discussed in *Land of Plants in Motion* as well as in his *Marathon Japan: Distance Racing and Civic Culture* (2015), *Parkscapes: Green Spaces in Modern Japan* (2011), *Radicals and Realists in the Japanese Nonverbal Arts: The Avant-Garde Rejection of Modernism* (2006), *Architects of Affluence: The Tsutsumi Family and the Seibu-Saison Enterprises in Twentieth-Century Japan* (1994), and *Artist and Patron in Postwar Japan: Dance, Music, Theater, and the Visual Arts, 1955–1980* (1982). His other book-length publications include *Global Indigeneities and the Environment* (coedited with Karen L. Thornber, 2016), *The Ambivalence of Nationalism: Modern Japan between East and West* (coedited with James W. White and Michio Umegaki, 1990), *Fire Across the Sea: The Vietnam War and Japan, 1965–1975* (1987), *The Historical Encyclopedia of World War II* (coauthor, 1980), *Valley of Darkness: The Japanese People and World War Two* (1978, 1986), *Farm and Nation in Modern Japan: Agrarian Nationalism, 1870–1940* (1974), and *Nishi Amane and Modern Japanese Thought* (1970). He has served as editor in chief of *The Journal of Asian Studies* and as a professor and director of the East Asian Library at the University of California, Berkeley.

Perspectives on the Global Past

Anand A. Yang and Kieko Matteson
SERIES EDITORS

Interactions: Transregional Perspectives on World History
Edited by Jerry H. Bentley, Renate Bridenthal, and Anand A. Yang

Contact and Exchange in the Ancient World
Edited by Victor H. Mair

Seascapes: Maritime Histories, Littoral Cultures, and Transoceanic Exchanges
Edited by Jerry H. Bentley, Renate Bridenthal, and Kären Wigen

Anthropology's Global Histories: The Ethnographic Frontier in German New Guinea, 1870–1935
Rainer F. Buschmann

Creating the New Man: From Enlightenment Ideals to Socialist Realities
Yinghong Cheng

Glamour in the Pacific: Cultural Internationalism and Race Politics in the Women's Pan-Pacific
Fiona Paisley

The Qing Opening to the Ocean: Chinese Maritime Policies, 1684–1757
Gang Zhao

Navigating the Spanish Lake: The Pacific in the Iberian World, 1521–1898
Rainer F. Buschmann, Edward R. Slack Jr., and James B. Tueller

Nomads as Agents of Cultural Change: The Mongols and Their Eurasian Predecessors
Edited by Reuven Amitai and Michal Biran

Sea Rovers, Silver, and Samurai: Maritime East Asia in Global History, 1550–1700
Edited by Tonio Andrade and Xing Hang

Exile in Colonial Asia: Kings, Convicts, Commemoration
Edited by Ronit Ricci

Burnt by the Sun: The Koreans of the Russian Far East
Jon K. Chang

Shipped but Not Sold: Material Culture and the Social Protocols of Trade during Yemen's Age of Coffee
Nancy Um

Encounters Old and New in World History: Essays Inspired by Jerry H. Bentley
Edited by Alan Karras and Laura J. Mitchell

At the Edge of the Nation: The Southern Kurils and the Search for Russia's National Identity
Paul B. Richardson

Liminality of the Japanese Empire: Border Crossings from Okinawa to Colonial Taiwan
Hiroko Matsuda

Sudden Appearances: The Mongol Turn in Commerce, Belief, and Art
Roxann Prazniak

A Power in the World: The Hawaiian Kingdom in Oceania
Lorenz Gonschor

Transcending Patterns: Silk Road Cultural and Artistic Interactions through Central Asian Textile Images
Mariachiara Gasparini

Land of Plants in Motion: Japanese Botany and the World
Thomas R. H. Havens